Staying on Top in Academia

Arne Pommerening

Staying on Top in Academia

A Primer for (Self-)Mentoring Young
Researchers in Natural and Life Sciences

 Springer

Arne Pommerening
Department of Forest Ecology and Management
Swedish University of Agricultural Sciences (SLU)
Umeå, Sweden

ISBN 978-3-030-65466-5   ISBN 978-3-030-65467-2   (eBook)
https://doi.org/10.1007/978-3-030-65467-2

Cover illustration: The photo on the front cover of this book, taken by the author in July 2016, shows a mature
Pinus radiata tree next to a hedgerow in the Roman fort Segontium at Caernarfon. North Wales, UK. A successful
academic, perhaps a professor, is like this large tree with the wide crown in the photo. Having been favoured by
lucky circumstances s/he has developed well and is now not only a blessing to the research community, but also
to many young and early career researchers who are like the small trees and shrubs growing up under the shelter
of the seasoned tree.

This Springer imprint is published by the registered company Springer Nature Switzerland AG
The registered company address is: Gewerbestrasse 11, 6330 Cham, Switzerland

What you have received more than others, e.g. in terms of skills, ability, success, a lucky childhood, harmonic domestic conditions, you cannot take for granted. All of this is the result of good will aimed at yourself. For the good you have received you have to pay something back in return by dedicating yourself to the well-being of others.

Albert Schweitzer, The Philosophy of Civilisation,
Volume II

Foreword

Educating young ambitious people is one of the most important activities in making our world a better place. During my own career, I had excellent mentors, and today, I myself am eager to help young talented people. Mentoring is important for learning about own strengths and weaknesses as well as for developing a personal view or style in research and teaching. Basically, this was the success of my own career and I am very grateful to my mentors for sharing their thoughts and experience so openly with me.

Today, I often see myself in a mentoring role of young researchers and find this highly rewarding. In a time where we have many opportunities, it is often difficult for young ambitious people to select the best options. Starting a successful career has a lot to do with hard work, being patient, but also with overcoming hurdles. Each career has its ups and downs and one of the important things is that young people need to learn that this is part of the process. These experiences make us stronger, if we draw the right conclusions. Often people ask what is important in mentoring people. For me it is trust and always being supportive and positive.

I congratulate Arne Pommerening on publishing this unique book. The open sharing of his own experience and thoughts is important to discover more about ourselves but also to appreciate the importance of mentoring young people for our society.

Wien   Hubert Hasenauer
December 2020

Preface

Since early on in my life I felt committed to Albert Schweitzer's epigraph at the beginning of this book. In a way the idea expressed there seems to have anticipated parts of the concept of academic mentoring and our duty to students and early career researchers, see Chapters 1 and 9. With this book I hope to pass on some of the blessings I have received from my family, teachers and mentors.

My interest in academic mentoring began at Bangor University (UK) where I was on the receiving end of research mentoring as a young, freshly recruited lecturer. In a less formal way, I had also received much support, encouragement and inspiration from quite a number of senior mentors who helped me develop my career. Many years later, I was rather spontaneously asked to give a seminar on academic career advice to doctoral students at Warsaw University of Life Sciences in January and February 2014. At the seminar, I was stunned by the great interest the doctoral students had in this improvised event and was delighted by their active participation. This was time well spent. Consequently, I decided to write down the things we discussed in the seminar, and over the years, I added to these notes.

At the seminar in Warsaw, I realised that academic career advice and mentoring in our rapidly changing scientific world was hardly offered and yet so much needed. Pushed to produce research results fast, academics sometimes tend to forget about taking time to counsel and mentor young researchers who are our future. Thankfully, since 2014, graduate schools and doctoral programmes in many countries have started to take these important topics on board and the first books on this subject have appeared in print.

Bearing in mind that time is precious, this book has deliberately been written as a primer covering the essentials to prompt reflection and discussion.

Complementing the table of contents, the subject index at the back of the book assists in finding specialised topics quickly. The text should appeal to master students, doctoral students, postdocs and other early career researchers. In addition, the book should prove useful for all who are interested in academic mentoring, both as mentees and mentors. The text may also be a valuable resource for academic tutors advising under- and postgraduate students on academic careers. As such, this book attempts to combine academic mentoring and career counselling in a brief, single volume.

The text is subdivided into two main parts. Chapter 1 introduces the concept of academic mentoring whilst Chaps. 2–10 discuss topics for mentoring conversations such as scientific thinking, doctoral studies, early career years, scientific storytelling, scientific presentations, teaching, research cooperation, job applications, behaviour and disappointments and basic data management.

Parts 1 and 2 can be read independently as can each single chapter. Research students and young researchers can use Chaps. 2–10 of the book for self-mentoring or for discussing some of the items with a mentor. Naturally, the recommendations made are meant as items to reflect on, that is, as "food for thought", rather than as firm suggestions. Many of them have worked well for myself and others, and the reader needs to carefully check them against their own objectives and research environment.

The book is based on my experience at different places in Europe, North America and China and on discussions I had with British, Estonian, Polish and Swedish doctoral students. I carried out academic mentoring as part of the SLU Graduate School in Applied Statistics and Scientific Computing that I had the good fortune to lead for 5 years. Recently, I have assisted Ann Grubbström from the SLU Educational Development Unit in Uppsala in organising a new academic mentoring programme for PhD students at the Swedish University of Agricultural Sciences. The discussions with Ann, Cecilia Almlöv and Katarina Billing have provided much incentive for writing this book.

I am grateful to Hendrik Heydecke, Mona N. Högberg, Irina Kuzyakova, Natalia Pommerening and Jonathan Pommerening for providing valuable comments on earlier drafts of this text. Christoph Kleinn kindly granted me office space and access to printers at his Chair of Forest Inventory and Remote Sensing (Göttingen University, Germany) during the challenging COVID-19 crisis. This was a great asset for this book project.

Hubert Hasenauer, Professor of Forest Ecosystem Management and currently Vice-Chancellor at BOKU University of Natural Resources and Life Sciences, Vienna, kindly wrote the foreword to this book. Hubert and I go back a long way, and I am glad we had the opportunity for an exchange on the topic of academic mentoring.

Umeå, Sweden   Arne Pommerening
December 2020

Contents

1

Empowering Researchers: Academic Mentoring

Abstract Academic mentoring provides guided discussion and fosters mentee reflection on individual educational and career path issues. Mentors usually are experienced, senior academics that informally help young researchers to tread the stony path of academia without personal agenda or compensation. Academic mentoring can range from activities providing assistance to overcome a single, specific problem to a long-term relationship of guidance. In this chapter, the concept and process of academic mentoring is explained.

Details of what can be discussed in mentoring sessions are then provided in the following chapters. Academic mentoring is supposed to unfold without much formalism and in accordance with the personal requirements of each research student or young faculty staff.

1.1 Introduction

Even if it may sound naïve it can be argued that it is the general purpose of academic life to make the world a better place through *research*, *education* and *outreach*, i.e. interaction with stakeholders and general society. Research gives us insight on the details of the system that we study and this knowledge can contribute to progress in, for example, technology, health care and nature conservation. At the same time we educate new cohorts of researchers, practitioners and policy makers who will, come the time, address important challenges of the future, when we are gone. Through reaching out we interact with the society that we are part of and explain the implications of our findings.

In this quest for making our world a better place, international academia has always strived to improve processes at universities and similar research institutions. In the past 30 years, PhD studies, for example, have been improved by introducing *supervisory committees*. Before this introduction research students occasionally encountered problems with their supervisor and this could often mean that they could not continue their studies. Along similar lines not every PhD supervisor was a gifted teacher and having a supervisory committee in place founded each PhD work on a more solid foundation so that problems and personal clashes could be solved more easily. A further improvement was the introduction of *individual study plans* that allowed to break the fuzzy concept of a PhD down into smaller, clearly defined parcels with milestones and deliverables. The introduction of *research courses* (with associated credits to collect) and *summer schools* that doctoral students now have to take at the beginning of their PhD time marked a further improvement with a view to better prepare research students for their research work by providing them with important knowledge and skills. In a similar way the requirement to give *talks at international conferences* during the course of a PhD and the need to write *cumulative PhD theses* based on a number of publications in peer-reviewed journals (see Sect. 3.3) have tremendously helped to improve the quality of doctoral theses and the employability of graduated PhD students. To give you a perspective, I still remember that my own line manager towards the end of the 1990s was surprised when I asked to attend an international conference during my postdoc time (!) and consequently declined the request. Since that time thankfully much has changed for the better. In addition *counselling services* at university campuses have also contributed to easing life and human interactions in academia. I am truly amazed by the improvements made in doctoral education over the last 20–30 years. Overall doctoral students have been granted more rights, security, freedom and support compared to earlier times. My own career would have been much more straightforward, if they were around when I was a PhD student.

Another such improvement of research education and continued research is *academic mentoring*. Despite lots of digital communication, pamphlets, brochures and open-day events there is still much confusion and disorientation among MSc as well as PhD students and junior faculty staff. What is the right direction to take, what should I specialise on, will there really be a job waiting for me in the end, if I go into all this trouble? What obstacles and conflicts do I need to put up with? For a beginner acting alone this is hard to answer, since s/he lacks the necessary insights and experience. This is complicated by the fact that many great changes have been made in academia over the last 10–15 years and the pace of change is rather fast.

There is, of course, line management provision, i.e. every student and young researcher has a supervisor whose role it is to provide guidance. Not very long ago, the terms "PhD supervisor" and "mentor" were synonyms. In some countries and depending on academic tradition, this connection between doctoral student and PhD supervisor used to be very strong and continued throughout the lifetime of the supervisor and even beyond. Famous senior researchers were often characterised as being the "student" of famous Professor So-and-so, although they already had many doctoral students themselves. Recently even academic ancestral lines were constructed, see, for example, the Wikipedia page for *academic genealogy*. Therefore the relationship between PhD student and PhD supervisor should not be underrated. It is clearly important.

However, as supervisors are often also concerned with annual performance reviews and other line-management duties, this kind of superior-co-worker relationship is not necessarily without tensions and it is not easy for a supervisor to act as mentor (Hopkins et al. 2020). Therefore research students find it awkward to approach their supervisors in a number of matters, e.g. in terms of the orientation of their research or job applications. Another example is that each academic is naturally enthusiastic about her or his research fields and when a research student enquires about the future prospects of this field, what is the supervisor likely to say about his or her own subject area other than praising it? Also, it may be awkward for a young researcher to discuss job applications with the supervisor or line manager, since this implies that s/he intends to leave the research group, which may sadden the supervisor. All such tensions, however small they may appear at the time, can eventually lead to greater problems. Cornér et al. (2017), for example, found that a lack of satisfaction with supervision and equality within the research community perceived by doctoral students were related to *burnouts*.

Particularly when things for whatever reason do not work well between young researchers and supervisors, the former often feel like falling between the cracks: It is hard for them to find someone they can approach and open up to without unintentionally upsetting the supervisor and thus worsening the relationship.

Unfortunately there are many factors that can inspire competition at work. Increasingly national research councils fund projects only in certain areas of science that are perceived as important, e.g. in climate change and health, whilst others receive little or no funding. This concentration of funding increases competition among research staff, because quite a few researchers are then employed who end up working in the same specialist field. This can even occur in one and the same research group. As competition among research staff is on the increase so are effects such as *bullying at work*, *sexual harassment* and *burnouts*. Therefore universities and similar research institutions are well advised to launch pre-emptive or even counteracting measures.

A possible solution is academic mentoring. In general, the term *mentorship* describes a relationship where a more experienced or a person with more knowledge (in a particular topic or situation), usually referred to as the *mentor*, helps to guide a less experienced or less knowledgeable person, commonly referred to as the *mentee* with the goal of developing the mentee by conversations between the pair. In its most basic meaning, mentoring is simply a conversation with a purpose (Hopkins et al. 2020). Mentoring is also characterised as *off-line help* by one person to another, where "off-line" implies that the line manager or superior of a mentee would not be involved. In fact, the ability to act as a confidant and for the relationship to be confidential is particularly important and as such means that the role of mentor should be separate from any role that involves evaluation or assessment. Combining these roles can alter the dynamic of the relationship and can make it less transparent, more guarded and so less likely to be successful (Biotechnology and Biological Sciences Research Council 2016). The mentoring relationship involves psychosocial support, career guidance, role modelling and communication. The goal of academic mentoring is to promote academic and personal development, particularly among new or recently appointed academic staff, by connecting them with others who can advise and guide.

Otis College of Art and Design in California (USA) have put forward the following definitions of academic mentoring:

> Academic mentoring provides *guided discussion* and *fosters mentee reflection* about individual educational and career path issues. Research shows that guided discussions of issues that impact mentees' *sense of control* over their academic outcomes enhance overall work success.

Academic mentoring can therefore contribute to increasing job satisfaction and inspires overall content and peacefulness at work. It also contributes to increased productivity and rate of publication and greater self-efficacy. A mentoring relationship has considerable value in helping individuals who are *in transition* to, for example, doctoral research, a new workplace etc. The one-to-one support that a mentor offers provides bespoke and tailored assistance, giving information when it is needed. As this support is less hierarchical than that offered by a line manager or supervisor, mentoring can feel more comfortable, if you are a mentee, allowing you to be more open and candid in your questions to your mentor without fear of judgement or assessment. Mentoring also helps new researchers integrate into their new research community. Mentees in fact stated that the mentoring scheme enhanced their productivity and alleviated isolation and loneliness (Hopkins et al. 2020). Morrison et al. (2014) found that medical staff at the University of Toronto (Canada) who engaged with mentoring were promoted on average a year earlier than those who did not.

> For mentors, mentoring is also a rewarding task, as this is a way to "pay back" for the good they have received from others during their career.

On the mentors' side mentoring can also lead to elevated research productivity and enhanced career satisfaction. Furthermore mentors go through a process of self-reflection and this may help them to become better supervisors and group leaders (Biotechnology and Biological Sciences Research Council 2016). Mentors reported feeling fulfilled and developing better interpersonal and communication skills, as well as personal skills such as compassion and patience while carrying out the role of mentor. Others reported benefits for mentors include greater commitment to and renewed enthusiasm for their work and more time to reflect on personal work practices. Yet again others successfully transferred mentoring techniques to their line management, to their private lives and there were also curious cases where mentors learnt new skills and technologies or cultural information from their mentees, something that is referred to as *reverse* or *upward mentoring* (Hopkins et al. 2020), see Sect. 1.7. For universities and research institutions, providing mentoring increases the chances of attracting and retaining the best people which will inevitably impact on metrics such as grant income and publications benefiting the institution as a whole.

To characterise mentoring better it is perhaps best to explain first what mentoring is not:

> - *Coaching* or any other in-house training,
> - A compulsory, formal course,
> - Parallel line management undermining authority,
> - An opportunity to evade uncomfortable decisions and tasks,
> - An alternative to supervisory committees or performance reviews,

(continued)

- A meeting with a trade union representative,
- A channel for making unjustified complaints and accusations.

Coaching is usually a narrow and functional approach of in-house training that familiarises new employees with the rules and procedures of a company or similar institution. Academic mentoring, however, has a much wider remit and is much more open-minded in its approach and objectives. In particular, academic mentoring should not be organised around or even tailored to the special system and set up of a particular university in a particular country, but rather aim at international academic careers in general and at generally accepted academic procedures. Using the obvious sports metaphor coaching mentees means using your own strategy and experience to drive them where you think they need to go. Mentoring people means showing them options and encouraging them to inform themselves and make their own decisions. This way mentees learn to see the bigger picture and gain confidence in their own judgement (Trinity Counselling 2020). Coaching is seen as a part of academic mentoring by some, I personally tend to disagree, since the strategies and methods involved are very different.

Also, academic mentoring typically does not come in the format of a formal course, e.g. a course for PhD students as part of a graduate school. In fact, academic mentoring and formality are mutually exclusive, because academic mentoring is strictly based on the freedom of thinking and expression.

A common misunderstanding among supervisors is that academic mentoring is a kind of additional line management interfering with existing line management. By contrast, academic mentoring is not part of any management and has no formal authority at all. Its strength is informality. As such the objectives of academic mentoring are very different from evaluative line management and supervision. Having said that it is important that the mentor-mentee relationship is carefully handled so that the line manager does not feel alienated or believes their authority is being challenged.

It would also be wrong for mentees to think that academic mentoring is a way to evade or avoid uncomfortable decisions and tasks, i.e. to secure the mentor's support in acting as an advocate to stop uncomfortable tasks. Whilst it is indeed the mentor's responsibility to help in cases of unfair treatment, academic mentoring should not be misused. Also, the mentor is not another member of the supervisory committee or a person charged with performance reviews or their corrections. The institutions of PhD supervisory committees and line management are not affected by academic mentoring.

In a similar way an academic mentor is not (and should not be) a trade union representative that you can call on in difficult cases. Trade union representatives have important roles to play, however, these differ from those of academic mentors. In many countries, you can, for example, ask trade union representatives to sit in with you and your line manager in performance review meetings, if you have reasons to believe that the meeting may otherwise be biased or opinionated.

Academic mentoring should also not be misunderstood as a channel for making one-sided complaints. Of course, discussing personal situations including negative elements is clearly part of the mentoring process, however, the mentor-mentee discussion always needs to be balanced and it is the mentor's responsibility to ensure that the dialogue does not drift off into a negative trajectory to remain there for the rest of the session.

> An academic mentor is an *independent facilitator* taking a *professional interest* in developing another person's career and well-being without personal agenda or ulterior motive. The mentoring process is about *sharing guidance, experience* and *expertise* (Trinity Counselling 2020).

The term 'mentor' originally derives from Homer's Odyssey where a man called Mentor is put in charge of Odysseus' son Telemachus when Odysseus leaves for the Trojan War (Hopkins et al. 2020). The mentor helps to set short-term and long-term goals relating to the personal development of the mentee, although it is the mentee who finally decides those goals. The mentor acts as a guide to help build personal and academic skills and to expand the vision of the mentee. S/he is also instrumental in navigating the inner workings of universities in general and of the university the mentee is based at. Using his or her own networks and sources, the mentor opens doors of opportunities for the mentee.

For example, as a mentor you may introduce your mentee to a successful grant writer for conversations about ways to approach writing a grant proposal or an academic with a strong publication record for advice on writing convincing journal papers (Hopkins et al. 2020). The mentor also facilitates networking and provides feedback on the mentee's academic work such as research and teaching. S/he helps others to meet their own goals and to identify their strengths and weaknesses as scientists and individuals (Gosling and Noordam 2011). It is not unimportant to emphasise that the mentor ideally should not be formally rewarded for her or his services so that the main

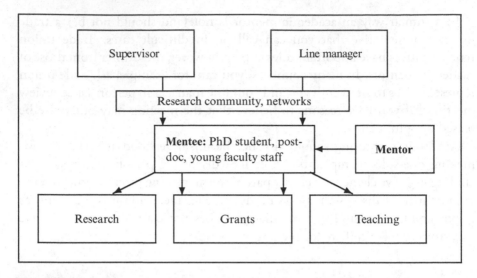

Fig. 1.1 Roles of supervisor, line manager and mentor in the academic development of a mentee

motivation for engaging in academic mentoring is the intention to help others. As such academic mentoring is carried out as an extra-curricular activity by research-active senior staff who are committed to supporting young colleagues.

Figure 1.1 summarises the roles of supervisor, line manager and mentor in the academic development of a mentee. In most institutions, the roles of supervisor and line manager are united in one person, but sometimes separate staff fulfil these roles: The supervisor advises in academic matters without having the formal authority of a line manager whilst the line manager makes management decisions and is responsible for the annual performance review without being involved in the details of the mentee's academic work.

The academic mentor usually comes in from the "sidelines", i.e. s/he is not involved in the day-to-day work or management of the mentee. The mentor does not necessarily have to be senior in terms of age or even career advancement, but can in some situations be just a few years or a few career steps ahead of the mentee. In contrast to the relationship with supervisor and/or line manager, the mentor-mentee relationship is informal. Such relationship can, of course, be "formally" organised by an institution (see Sect. 1.6), however, if it is to work properly the relationship always has to be of a voluntary and informal nature. The majority of mentor-mentee meetings are related to the three main academic activities, i.e. research (including research talks at conferences etc.), grant capture and teaching.

Some guidelines recommend that each mentee has a range of different mentors to discuss a variety of different topics (Biotechnology and Biological Sciences Research Council 2016). It is certainly a good idea to ask several people independently for advice, particularly when important questions are concerned. However, it may logistically be difficult to secure a range of mentors given that time is in short supply.

1.2 Mentor-Mentee Partnership

As discussed before, academic mentoring is a process in its own right independent of other processes such as line management and promotion procedures. The mentor-mentee relationship clearly is an informal partnership where both partners interact eye to eye on equal terms. For this to happen well it is best that the mentor is not part of the same research group as the mentee. In most cases the mentor may even come from a different department but can also be employed by a different university or research institution. When selecting mentors, be mindful of possible rivalries or politics between institutions or individuals so that they do not potentially get in the way of the mentoring. Seeking, for example, guidance from your supervisor's greatest competitor could call for trouble (Gosling and Noordam 2011). Asking emeritus staff to help may also be a good choice, since these have lots of experience, sufficient distance to the day-to-day business of the university and are often still involved in research. In ideal terms the mentee chooses her or his mentor in the same way as all initiatives in the partnership should be made by the mentee. A particular mentor may be sought because they have some technical expertise or professional knowledge or they are a source of organisational knowledge. However, if, for example, whole universities launch an official mentoring programme (see Sect. 1.6), administrative constraints and efficiency often require a different approach. In such cases a *matching* service is offered, where mentees can state preferred properties of suitable mentors (including gender, cultural background, research field, home department) or suggest a person they know. If such matching services are implemented, mentees should have the opportunity to freely change mentors at any time with no strings attached. If there is no mentoring programme at your university or research institution, approach a suitable mentor yourself and ask this person, if s/he is willing to act as your mentor.

An important property of the mentor-mentee partnership is that the mentee "*owns the process*" (Trinity Counselling 2020). This implies that all initiatives are typically made by the mentee, i.e. the mentee calls for meetings and decides the agenda of the meeting, the items to be discussed. As part of the process the mentor guarantees the strictest confidence at the discretion of both parties. The mentor is supposed to set an ethical example and to exhibit integrity and professionalism. As such the mentor inspires ethical behaviour in the mentee and acts as a positive role model. Mentors also frequently act as role models to their mentees, because they are often in a position to which a mentee is seeking to aspire (Hopkins et al. 2020). In situations where the mentee reports a conflict, the mentor suggests peaceful solutions that are of benefit to all parties involved. You can choose a mentor for his or her career choices that you admire, for a great work-life balance or for the high-ranking publications the mentor has achieved. There are certainly more criteria that can be applied but another one that clearly is essential is a certain wisdom that comes from the heart (Gosling and Noordam 2011).

Typically the mentor does not get directly involved in managing conflicts. S/he advises the mentee and suggests solutions. The mentee then decides which of them to take forward and tries to solve the problem at hand using the chosen option. Commonly, there is no situation where the mentor has a word with the mentee's line manager or supervisor, nor does s/he talk to any other parties, e.g. to counselling services, about a mentee's problems. However, there can be situations of emergency, where, for example, unlawful or even criminal offences against the mentee or a third person are involved, and the emergency of the situation suggests a direct and immediate action by the mentor, see Sect. 1.6. However, even in such situations the mentor can in principle only act, if the mentee explicitly authorises the mentor to do so.

Naturally no mentor can know everything or help in every situation. It is perfectly fine and in fact an important role of mentors to refer mentees to other, specialised university services or to colleagues who may be better placed to help.

1.3 Mentoring Focus

What individual mentoring sessions are about, is typically decided by the mentee and information on the items addressed should not be disclosed to anyone unless both parties agree to do so under very special and clearly defined circumstances. The key activity is to aid mentees to talk through their own thoughts and decisions regarding issues they may face with a view to encourage self-reflection (Biotechnology and Biological Sciences Research Council 2016). Mentors often are in a position to motivate the mentee to achieve their goals, but they also should have the ability to challenge assumptions and encourage different ways of thinking. It should also be part of academic mentoring that the process fosters a sense of belonging to the university community and pride in the university. Topics usually offer themselves and come up as questions or challenges in day-to-day work and dealing with other people. Here is an expandable list of possible topics in no particular order:

- Settling in (getting accustomed to a new country, town, institution).
- Time management,
- Training and career development,
- Develop intra- and interdisciplinary inter/national networks,
- Devising scientific ideas and research visions,
- Publication pipeline,
- Targeting external funding,
- Teaching and research portfolio,
- Balancing/prioritising academic tasks,
- Work-life balance,
- Editorial experience,
- Research trends and frontiers,
- Taking opportunities for extra-curricular activities,
- Team-building and interpersonal skills,
- Professional communication,
- Finding jobs and job applications (including cv and profile reviews),
- Practising job interviews,
- Ethics and role model,
- Supervisor relationship,
- Conflict management.

It depends on both mentee and mentor how far the latter gets involved in the former's papers, research proposals, conference applications etc. Some mentees prefer a hands-on approach where the mentor actively looks at certain documents and gives detailed advice, others prefer discussions. Obviously this also depends on the time constraints of both mentor and mentee. However, in principle anything is possible and the two partners are totally free to design their mentor-mentee relationship as it works best for them.

In my experience, in the majority of cases mentoring meetings are about career development. Some mentees seek orientation by asking about the meaning of science and research, see Chap. 2. Others wondered whether an academic career or a career outside the university may be best for them. Cases of conflict management are also not uncommon. I also had situations where the mentees asked for advice on proposal writing and I commented on the first proposal drafts. The topics listed above can also be mentored in specialised mentoring programmes or with specialised mentors, see Sect. 1.7, however, the most common situation is a general mentoring scheme where mentee and mentor discuss different topics as the need arises.

Naturally, if a mentee does not have anything particular to discuss, s/he can cancel a scheduled meeting or not call for one in the first place. It is also possible just to meet for catching up over a cup of tea or coffee without a particular agenda and then to see how the conversation takes its natural courses.

1.4   Place and Time

Once the mentee has contacted the mentor for a meeting it usually falls to the mentor to arrange a suitable location and time. It helps to book a pleasant, well-maintained meeting room with a transparent door or wall, so that the mentee does not feel too isolated or locked up. It is good practice to place a table between mentor and mentee to give both sufficient space and freedom. Always maintain clear boundaries and behaviours that ensure a safe, productive relationship between mentors and mentees. Any uncomfortable situation should be avoided (Trinity Counselling 2020). Sometimes it can be a good idea to have a meeting off campus, e.g. in a café. It may also work well to sit outside in a public park or to go for a walk. Everybody is different and you may want to experiment a little. The situation in an environment different from work sometimes helps to unlock people's minds. Here it is just important not to select a lonely location or a place where other people sit close and can easily eavesdrop and overhear the conversation. Aim for a time when both, mentor and mentee, are not too tired. If the mentee thinks it would help, the mentee can also ask a friend or independent witness to be present in the meeting. It almost goes without saying that during a mentoring meeting, mobile phones and other distracting devices should be turned off. Laptops and similar devices are, of course, sometimes necessary to discuss documents or to look at websites together, however, for obvious reasons emails and other communication should not be checked during mentoring sessions.

Some mentees prefer regular meetings for an extended period. Others are content with only a few meetings or even with a one-off. Sufficient time should be allowed between successive meetings for both parties to reflect on what has been said and assess how well the process is going. Any programme that your personal mentor-mentee relationship is part of should allow any of these options. After all an important goal of academic mentoring is that the mentee develops *independence* through professional guidance rather than becoming dependent on the mentor for constant support (Gosling and Noordam 2011).

Although face-to-face mentoring meetings usually work best, telephone or video conferencing calls as well as occasional mentoring through emails are also possible.

1.5   Mentoring Conversation

Key mentoring principles include *creating space for reflection, fostering mentee confidence, fostering independence* and *role modelling* (Hopkins et al. 2020; Trinity Counselling 2020). Mentoring conversations should be guided by these principles and there are a few useful techniques that support them. Actively providing advice and building confidence and initiative is one part of a mentor's role. Other typical roles include being a sounding board for the mentees' various ideas or a facilitator. A crucial ability here is active listening without prejudice or emotions, where the mentor allows for sufficient time for the mentees to say what they have to say and to observe their body language (Trinity Counselling 2020).

Active listening is key to any mentoring. Being fully present in the conversation and actively listening to your mentees will contribute to maximising the value of the meetings. While listening use short words of encouragement (like 'yes', 'right' or 'uh-huh'), nod or smile to keep the mentee talking. After listening carefully it is a good idea for the mentor to reflect on, to paraphrase and to summarise what s/he has heard. This will help the mentor to develop a better understanding of the problem and encourages the mentee's self-reflection (Hopkins et al. 2020). These techniques ensure you do not miss anything that the mentee is saying and give them the chance to correct how you interpret them. Also allow space and time for thinking (Trinity Counselling 2020).

Reflecting in this context implies mirroring the meaning and feeling of what somebody has said. Usually this involves repeating the last few words of what the mentee said. This communicates that you have heard and are absorbing what is being said. Here is an example (Trinity Counselling 2020):

> Mentee: "I won't be able to do it. I'll just get up and freeze!"
>
> Mentor: "You'll freeze?"

Paraphrasing allows a speaker to re-hear a statement and thus verify that the listener did, in fact, listen to them. The technique also invites the mentee to explore further or to understand better whatever is being discussed. The paraphrasing can be introduced using phrases such as "Am I right in thinking …", "I hear you saying that …", "If I understand correctly, you …" or "It sounds like you …". They give the mentee the opportunity to confirm or reject a paraphrase and the mentor can correct any perception error on his part. The mentor should not judge, dismiss or use sarcasm. S/he should also not add to what the mentee has said and avoid interpretations (Trinity Counselling 2020).

The mentor is not expected to provide instant answers but to guide the mentee towards the right answer for him or her. A crucial element of mentoring conversations is that the mentor asks the mentee open questions after s/he has stated the problem he or she is concerned with. Question asking empowers the mentee to find solutions to their own issues by reflecting on them within the safety of the mentoring partnership. Open questions are those that cannot be answered with a single word. They usually begin with 'what', 'how', 'tell me' or 'why' and elicit answers of more than a single word (Hopkins et al. 2020).

Another useful technique is the 'slap sandwich' or 'hamburger technique'. Each piece of criticism or uncomfortable suggestion is softened by positive comments and is therefore easier to hear and to accept.

It is important the mentor does not offer solutions straight away but rather helps the mentee to find solutions themselves to the particular challenge they have identified. It is much more empowering to help someone to reach a decision themselves, even if it takes longer and even if the mentor disagrees with the mentee's decision. The mentor should certainly share any hints and insights and rather provide options for the mentee to consider instead of giving instructions. Mentors should allow mentees to reach their own conclusions in their own time (Hopkins et al. 2020; Trinity Counselling 2020).

1.6   Setting Up a Mentoring Programme

From the point of view of an individual seeking the help of a mentor, there is, of course, no need for an official mentoring programme to exist. The research student or young researcher in question simply has to approach a mentor of her or his choice and can then arrange meetings with this person. It is very likely that in each research institution or at least somewhere nearby there are sufficient people around that are happy to make their time and experience available to young colleagues. Ask around, your colleagues may know an academic who has the experience you are interested in or you can get information from institutional and web-based online profiles about academic experience. You could speak to peers and academics that you meet at seminars and similar events. Ideally such a person should not be in your research group or share your office to be able to give you a different perspective (Hopkins et al. 2020). You can also mention your interest in a mentoring relationship in your annual performance review. This voluntary and mentee-driven process is in fact the best way mentoring can be set up.

In some cases, however, a university or research institution wishes to organise an official mentoring programme. There can be a number of reasons for this:

- To raise awareness among university managers,
- To encourage as many researchers as possible to take advantage of mentoring,
- To address problems in the organisation, e.g. gender issues and sexual harassment,
- To complement doctoral programmes and graduate schools,
- For proactive conflict management throughout the university.

Such programmes can prove to be key mechanisms through which research organisations can support the professional development of their staff, particularly of young, newly appointed researchers (Biotechnology and Biological Sciences Research Council 2016). For organising and rolling out mentoring programmes, Trinity College Dublin (Ireland) and the Swedish University of Agricultural Sciences (SLU, Sweden) have adopted a process as outlined in Fig. 1.2.

It is useful to have some recognition of the programme from senior staff members, for example, an *endorsement* from the Vice-Chancellor or heads of departments. Such endorsement helps potential mentors to feel that the time

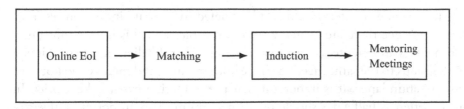

Fig. 1.2 Description of a possible way to organise and roll out a mentoring programme at a university or other research institution. EoI: Expression of Interest

they offer is permissible and viewed as valuable by the institution (Hopkins et al. 2020).

The mentoring programme is launched on a website where both mentors and mentees can sign up by submitting an expression of interest to act as mentors or mentees. Along with personal information they provide information on background, subject area, skills, academic and work history, interests, hobbies and goals that can be used in the matching process. For example, international researchers adapting to a new country can then be matched with mentors from their own culture to help them to settle into the new environment. When mentees apply, they should be asked to outline what they would like in a mentor and any specific requirements that would help you match their needs to one of the individuals in your pool of mentors (Hopkins et al. 2020). Only those employees should sign up that really want to enter a mentor-mentee relationship, believe in its value and are willing and able to commit sufficient time to it. The matching is done by dedicated and experienced staff in the university administration and can be changed later should the need arise. Hopkins et al. (2020) recommended that the staff responsible for matching make an effort to get to know the mentors personally so that their personal knowledge of the mentors can aid in the matching process.

Matching is followed by a short briefing session. This is an induction event where all mentors and mentees are invited and are briefed about the *code of conduct* (see the following boxes), objectives of the programme, roles and responsibilities. For ensuring confidentiality Hopkins et al. (2020) gave an example of a generic safeguarding statement that could be part of the code of conduct document:

> The conversations we have will be confidential and information will only be disclosed with your consent or in exceptional circumstances where your health, safety and well-being or the health, safety and well-being of others is a concern.

Mentors and mentees could also be briefed in separate induction sessions, however, since there are so many common topics and this is an opportunity for mentors and mentees to get to know each other, I believe it is best to invite both groups to the same meeting. Since both mentors and mentees do not have time in abundance, it is important that the induction event is kept brief. In my experience, half a day should be the maximum. Also, important university services such as student finance, accommodation, well-being could be briefly reviewed so that the mentors can signpost their mentees to appropriate help.

The event can also include hands-on activities that are useful for the mentoring process, such as allowing for time for matched mentors and mentees to get to know each other or a *"speed-dating session"* where everybody randomly talks to another person from the induction crowd for 2–5 minutes about what they expect from the programme. Based on Hopkins et al. (2020) the briefing could include the following items:

- Explain the nature of mentoring,
- Show the mentors and mentees the process of how the programme works,
- Review university services available to your mentees,
- Demonstrate a few mentoring techniques,
- Introduce the code of conduct establishing what mentors and mentees can and cannot do.

For bringing across any of these items videos, activities and discussions can be used to break up the rigidity of formal presentations. A variant of the induction session can also be offered as a distance-learning course, e.g. in the format of a video (made from a previous induction event) or as a write-up. At more advanced stages of the mentoring programme only newcomer mentor and mentees are invited to the induction session.

After participating in the induction event mentors and mentees independently arrange meetings and the real mentoring starts now. Some mentoring schemes recommend that both parties sign a *mentoring agreement* in their first meeting to avoid potential misunderstanding later on. This agreement can detail topics to be discussed, confidentiality, contact arrangements, frequency, timing and locations of meetings and the end of the formal mentor-mentee relationship among other things. However, a formality like this may already violate the voluntary and informal character of academic mentoring and some of these aspects may have already been covered by the aforementioned code of conduct.

To break the ice, a topic for the first meeting can be suggested, e.g. career development, but it is best, if the mentees freely chose according to their inclinations and requirements. An informal start could include encouraging mentees to talk a little about themselves followed by sharing a little of the mentor's academic journey (Hopkins et al. 2020).

In the case of problems they cannot solve themselves, the team who prepared and launched the mentoring programme can usually be contacted for issues that may occur in individual mentor-mentee pairs. Some universities offer a formal *mentoring certificate* to mentees on the completion of the mentoring programme. Personally, I believe that this is too formal an approach and defeats the original intention of mentoring, however, there may be contexts where this is appropriate. If in doubt, the self-perpetuating processes in mentoring programmes should not be disturbed by too many formalities including forms, reports or reviews.

Throughout the process, feedback can be collected from both mentors and mentees with a view to improve the mentoring programme, but again this should be done in moderation so that the participants do not feel burdened with paperwork. There is also occasional advice to require the mentee to send a summary of what was discussed and agreed to the mentor after each meeting. Whilst such paper trails can be useful in some situations, this requirement may be too rigid and oppressive in others.

Software packages are currently being developed to support mentoring programmes, see, for example, SUMAC on https://sumacmentoring.co.uk.

1.7   Specialised Types of Mentoring Programmes

In many cases, the organisers of mentoring programmes devised specialised themes for them that – as a consequence of the specialisation – limit mentee participation. This may be for practical reasons as a consequence of limited resources, but sometimes universities and similar research organisations feel they have to support particular groups of students or employees who are perceived to be disadvantaged. For example, the provision of mentoring for women and minority groups is often perceived to be beneficial, as the loss from academia of talented researchers from these groups is especially high and mentoring may help to address this issue (Biotechnology and Biological Sciences Research Council 2016).

Göttingen University (Germany), for example, runs a specialised mentoring programme for female postdocs and another one targeting PhD students and postdocs considering to work in industry after graduation or after the current contract with the university has ended. The university also runs a specialised mentoring programme for researchers who are interested in working in science management. Jena University (Germany) and Freiburg University (Germany) decided to organise a specialised mentoring programme dedicated to international students. Here mentors and mentees need to be sensitive to potential misinterpretation in language and behaviour particularly where there are differences in gender and/or culture.

Trinity College Dublin (Ireland) launched an interesting alumni-to-student career mentoring, where former graduates of Trinity College with interest in sharing their professional experience support current Trinity students to help them bridge the gap between university education and professional career interests. The same university also maintains a *momentum mentoring* aiming at continued mentoring of existing staff. Swansea University (UK) offer a mentoring scheme to support academic promotion.

The University of Natural Resources and Life Sciences (BOKU) at Vienna in Austria maintains a specialised programme for mentoring women in forestry. The same university has also organised an alumni-to-student mentoring similar to that at Trinity College Dublin. The mentoring of students in Vienna is jointly organised by all five universities of the capital.

Similar to other universities, Dundee University (UK) has organised an early-career mentoring scheme, while the Swedish University of Agricultural Sciences (SLU, Sweden) currently focusses on doctoral students (see Chap. 3). Early career researchers (ECR) are often the target group of mentoring programmes, as they are at a pivotal career stage and therefore require mentoring support. They are frequently on fixed-term contracts with a dedication to their research projects but also have to focus on securing the next position. ECRs are usually the main drivers of research outputs within research teams and are the future of academia, who will bring in research funding and attract students. At the same time they often start a family (Hopkins et al. 2020), see Chap. 4.

Hopkins et al. (2020) also discussed mentoring for the transition out of academia. This can relate, for example, to doctoral students who wish to work outside academia once they have completed their PhD programme. This focus can also relate to mentoring programmes such as Göttingen University's mentoring for science management. Interestingly, here instead of offering dedicated courses, a mentoring programme has been chosen for educating staff. The alumni-to-student career mentoring offered by Trinity College Dublin and BOKU University Vienna is partly also mentoring for the transition out of academia.

Some institutions have launched so-called *buddy schemes*. These peer support programmes usually are smaller programmes compared to mentoring programmes and are designed for particular purposes such as adjusting into life at a particular university and in the corresponding university town. It is also possible to establish writing buddy schemes where two authors (who do not publish together) support each other and, for example, set writing targets for each other among other lines of support and encouragement (Hopkins et al. 2020).

The aforementioned reverse, upward or mutual mentoring is a mentoring type where older, senior or more experienced staff are mentored by younger, more junior colleagues or students. As such it turns the conventional model of mentoring on its head. This type of mentoring takes advantage of the skills and experience of young mentors and is most commonly used to support the older mentee in developing competences related to emerging technologies, IT, social media or cultural backgrounds that their younger colleagues often take for granted (Hopkins et al. 2020). Reverse mentoring can develop naturally in an existing mentor-mentee relationship but can also be organised as a specialised mentoring programme. Particularly in the latter case, mentoring is mutual and reciprocal and leads to shared benefits. The breakdown of the usual perceived hierarchical barriers between mentor and mentee facilitating freer and more open dialogue is very beneficial. At the same time reverse mentoring is likely to enhance a deeper understanding, empathy and respect for each other (Hopkins et al. 2020).

Naturally it is possible for a university to run several mentoring programmes at the same time. These typically operate across departments or faculties, i.e. they are usually not bound to particular subject areas. It is important to include existing long-term staff like in the case of the momentum mentoring scheme of Trinity College Dublin. It may also be useful to start with a pilot programme in a single faculty or staff group with a small number of mentor-mentee pairs. Once this has turned out to be a success you can plan for larger schemes. If a prospective mentee does not feel covered by the existing mentoring provision, it is always possible to organise an individual mentor-mentee relationship or to join an existing mentoring programme and tell the staff in charge that individual objectives slightly differ.

1.8   Group Mentoring

Many British universities traditionally have a *tutoring system* in place. As part of this, every year a number of new under- (BSc) and postgraduate (MSc) students are assigned to one academic, i.e. each student has a designated personal tutor. In contrast to academic mentoring, students usually cannot select their mentors and there is no matching process, but rather a simple allocation. As a result each academic tutor has a group of 5–30 tutees from different study years to take care of. Academic tutoring is particularly instrumental in helping students to successfully achieve the transition from school to university (Molina Jordá 2013). The responsibilities of tutors share many similarities with those of academic mentors and include pastoral care, each term aiding the students' module selection, organising courses and discussing life in the university town. The role of inviting to tutor-tutee meetings here rests with the tutor and it is possible to invite mentees to group meetings which all mentees of a certain tutor attend, but most common are 1 : 1 meetings. Sometimes tutors are also asked to formally teach key skills to tutees, e.g. how to give a presentation, to write an essay or a summary, scientific ethics and plagiarism. Such teaching is typically done as part of group mentoring, i.e. each academic asks all their tutees to attend such meetings and at the end the learning success is formally assessed. At some universities tutors also help to arrange student internships as part of so-called sandwich years, i.e. a practical year usually between second and third study years of BSc degrees, and visit the students during their internships to discuss their progress.

Academic tutoring is of great value to both students and mentors and proactively helps reduce the number of student drop-outs. In times of crisis, e.g. when pandemics make student life hard to cope with, academic tutoring is often of crucial importance. Universities and similar institutions could consider introducing academic tutoring at the same time as they launch academic mentoring so that undergraduate students are covered as well. It can also be considered, whether certain elements of mentoring can also be carried out as a group effort. Cornér et al. (2017), for example, recommended collective forms of supervision of doctoral students so that they can develop national and international networks. The seminar on academic career advice I gave at Warsaw University of Life Sciences in 2014 (see Preface) can be considered a group-mentoring effort. Another time I was invited to a coffee break organised by doctoral students and they informally interviewed me about my experience with various academic issues and particularly with devising interesting research questions. This was also a variant of group mentoring.

2

The Shape of Science

Abstract What sets researchers and scientists apart from the rest of society? Do you need to be a nerd or a kind of weirdo to become a scientist or perhaps a workaholic? Do scientists and researchers think and behave in ways different from the non-academic population?

The simple truth is that most researchers are indistinguishable from anybody else you might meet on the street or anywhere else in public life. True, some scientists are a little eccentric but this is only a tiny minority. Most are just like you and me. What is perhaps different is that they adopt a special way of thinking when they are at work. They also apply a specific scientific framework when dealing with research questions and problems. In this chapter, we will explore this way of thinking and the general methods to set the scene for other chapters of this book.

2.1 What Scientists Do

Scientists typically observe and measure the world around them. They gather data based on their observations and when they think they have enough information to answer the questions they have asked, they try to make sense of what it all means (Gosling and Noordam 2011).

In this process, most scientists use a *reductionist* approach. Scientific reductionism involves the description of phenomena in terms of simple or causal hypotheses (Kimmins 2004). In order to make sense of the complicated

© The Author(s), under exclusive license to Springer Nature Switzerland AG 2021   **23**
A. Pommerening, *Staying on Top in Academia*,
https://doi.org/10.1007/978-3-030-65467-2_2

phenomena they study, each researcher must break down the particular problem into small components:

- Observation,
- Constructing a hypothesis,
- Carrying out experiments to test the hypothesis,
- Arriving at conclusions,
- Formulating a theory.

These steps commonly constitute the scientific method. If carried out correctly, the ultimate goal of the scientific method is to construct an accurate representation of the physical world (Gosling and Noordam 2011). What can happen, if mistakes are made or the scientific method is not taken seriously, has recently been described in detail by Ritchie (2020). In forest ecology, Newton (2007) distinguished between *research*, *survey* and *monitoring* in scientific studies:

- *Research* is generally undertaken to answer a specific question or to test a hypothesis,
- A *survey* is typically a descriptive piece of work, which might be more open-ended in nature than a research project and might not have such a clear outcome. This often also includes studies on data that were not collected from statistically designed experiments such as data from *observational field plots*,
- *Monitoring* is a form of survey that is designed to be repeated over time, enabling trends in some variable of interest to be determined.

The international scientific community tends to place greater emphasis on research rather than on survey and monitoring work and this is reflected in the content of scientific journals. Research work is believed to have a clear relationship to relevant theory, however, surveys are important parts of research programmes and the three types of studies naturally complement each other (Newton 2007). In this context, it is also useful to distinguish between *fact*, *theory* and *hypothesis* (Gosling and Noordam 2011):

Fact: Something that is known to be true, e.g. fire burns wood into ash, water is solid at temperatures below zero °C.

(continued)

Theory: A conceptual framework that is believed to be true and can be used to explain observations and to predict new ones. For example, the path the sun follows as it crosses the sky can be explained by the theory of gravity. A theory, even if its not believed to be true, can also be used as a basis for argument (Porkress 2004).

Hypothesis: A working assumption: A hypothesis can be defined as a prediction based on some explanation of the observations made (Newton 2007). Usually this assumption is formulated *before* experiments are carried out to test it. If the hypothesis holds up against existing and newly obtained data, the scientist may formulate it as a theory.

Once a hypothesis has been formulated, it can potentially be tested using statistics. *Statistical hypotheses* often refer to the values of parameters of the parent population from which the sample of data has been drawn, e.g. mean and variance (Porkress 2004). This process of predicting an outcome by deducting what is logically consistent with a hypothesis, followed by its testing against observations made, is known as the *hypothetico-deductive* scientific method. Statements of objectives are often intertwined with hypotheses (Newton 2007). Statistical tests usually include a so-called *null hypothesis*, which commonly states that "nothing has happened". This is a conservative statement and we are usually interested in the reverse. Thus we try to falsify or reject the null hypothesis using our data and appropriate statistical tests. The results obtained from experimentation are used inductively to support or reject the initial theory or to derive alternative theories. If after many tests one fails to reject a theory, it may be elevated to a *scientific principle*. After many years of evaluation such a principle may be accepted as a *scientific law* (Kimmins 2004). Biases aside, this method is the best approach we have to accurately answer the questions about the physical world in which we live. Occasionally one can make errors using the scientific method. The most common ones are listed below (Gosling and Noordam 2011).

1. *Not proving a hypothesis*
   This involves mistaking the hypothesis for an explanation of a phenomenon without having performed any tests to verify the hypothesis. Sometimes what we think of as common sense, logic or intuition tempts us into believing that no experimental proof is necessary to prove the hypothesis, because the answer seems so obvious from the start.

(continued)

2. *Discounting data that do not support the hypothesis*
   In the ideal situation, a scientist is open to the possibility that the hypothesis is either correct or incorrect. If, for example, a researcher has a strong belief that the hypothesis is true or false before collecting any data, there may be a psychological tendency to find something 'wrong' with any data that does not support the researcher's expectation (Ritchie 2020). This error is sometimes also related to the fact that *negative results* are difficult to publish. Negative results are those that do not support the hypothesis and therefore nullify the aim of research.
3. *Over- or underestimation of systematic errors*
   Many discoveries were missed by researchers whose data pointed to a new phenomenon, but the data were mistakenly attributed to 'experimental noise'. Conversely, data that are part of the normal variation of the experimental process were taken as evidence for a new discovery.

Statistics can clearly help avoid some of these mistakes but also open communication among members of a scientific field in the form of publications and conferences. In this way, the biases of individuals will most likely be cancelled out as other scientists try to reproduce their results. In time, a consensus may develop in the research community as to which publications have withstood the test of time (Gosling and Noordam 2011).

In principle it is our duty as scientists to impartially publish all data, no matter what the outcome, because a negative finding is still an important finding and in the spirit of transparency we are supposed to present all sides of a story. Negative findings are in fact a valuable component of the scientific literature (Matosin et al. 2014; Ritchie 2020). Increasingly, there is pressure on scientists to aim at high-impact results. This forces many to tuck away negative findings, i.e. non-significant results (those that support the null hypothesis) and to focus on their positive outcomes. Fanelli (2010) even concluded that papers are less likely to be published and to be cited, if they report negative results. The fact that negative research results are largely excluded from publication can lead to a publication bias (Ritchie 2020). For example, when a certain popular system of methods does not result in significant results and this goes unpublished, other research groups may continue to follow the same lines of thought and paths of investigation with the same outcome, ultimately wasting time and resources. This has made many researchers wary and some have engaged in creating new journals that publish the 'rejects' (e.g. *Journal of Negative Results in Biomedicine, The All Results Journal, Journal of Articles in Support of the Null Hypothesis* etc.). Negative findings can have positive outcomes and positive results do not necessarily equate to productive science (Matosin et al. 2014).

2.2   Experimental Design

The statement "You can observe a lot just by watching." is certainly true, however, it clearly has limitations. Friedrich Wilhelm Leopold Pfeil (1783–1859), an eminent representative of early forestry in Germany, was famous for his motto "Ask the trees!" which stressed watching and observing. Naturally observations often lead to hypotheses and theories, but they are not sufficient when you want to understand what happens to a process with varying factors. For understanding cause-and-effect relationships there is a need to change input variables and study the corresponding changes in the output.

Only in very few cases processes are so well understood that a law (see end of Sect. 2.1) can be formulated, for example, Ohm's law

$$I = \frac{V}{R}, \tag{2.1}$$

stating that electric current, I, is proportional to voltage, V, and inversely proportional to resistance, R (Montgomery 2013). In most other cases experimentation is an important scientific method. The results reveal statistical relationships hinting at causal relationships.

> An experiment is a test or a series of test runs in which purposeful changes are made to the input variables of a process or system with a view to identify the reasons for changes in the output response.

Experiments are used in all scientific fields to study a particular process or a system. We are particularly interested in learning which input variables or combination of input variables are responsible for changes in the output variables. Later we may want to develop a model relating the response to the main input variables (Montgomery 2013), see Sect. 2.3.

Here *factors* are explanatory variables that are most influential on the output of a process. The main problem-solving strategies and strategies of experimentation can be summarised as follows (Montgomery 2013):

1. *Trial and error*
   This is a fundamental method of problem solving which has proved very successful in human evolution. It is characterised by repeated, varied, but arbitrary attempts of selecting factors which are continued until success or until a person stops trying. Trial and error does not employ experimental design and associated statistics. Often applied by practitioners, e.g. in cooking, gardening, agriculture and forestry.
2. *Best guess*
   Take an educated guess in terms of selecting a combination of factors based on prior knowledge (literature) or experience and see what happens.
3. *One factor at a time (OFAT)*
   Vary each factor separately over its range with the other factors held constant at the baseline level.
4. *Factorial experiment*
   OFAT fails to consider any possible interaction between factors. In factorial experiments, factors are varied together to include interactions. Interactions occur, if the differences in the response variable depend on the variation in two or more factors.

Experiments cannot only be organised in the field and the lab, but also at the computer. A model (see Sect. 2.3) can be used to run a number of simulations with varying model parameters or other inputs and then the results can be analysed in exactly the same way as those from field or lab experiments.

Montgomery (2013, p. 14) described the iterative workflow of scientific experimentation as follows implying that individual steps or the whole procedure may need to be repeated a number of times:

1. Recognition of and statement of the problem,
2. Selection of the response variable,
3. Choice of factor, variants of factors (levels) and factor ranges,
4. Choice of experimental design,
5. Performing the experiment,
6. Statistical analysis of the data,
7. Conclusions and recommendations.

Steps 1 and 2 are strictly speaking part of pre-experimental planning and steps 2 and 3 are often carried out at the same time (Montgomery 2013). The purpose of the statistical analysis is to discriminate between scientifically interesting variation and background "noise". We achieve statistical significance, if our results show more variation than we expect to occur by chance alone (Crawley 2005).

In contrast to monitoring, scientific experiments are more "controlled" and need to follow a number of important principles of experimental design. Following these design principles is a necessary pre-requisite for statistical analyses:

- Randomisation (must),
- Replications (must),
- Blocks (can),
- Controls (recommended).

Randomisation implies that both the allocation of experimental units, e.g. plots, and the order in which the individual runs are to be performed are randomly determined. Statistical methods require the observations (or errors) to be independently distributed random variables and randomisation usually makes this assumption valid. By properly randomising the experiment, we also assist in "averaging out" the effects of external factors that may be present and this avoids systematic errors (Montgomery 2013).

Replications are independent experimental runs. Every experiment should have replications of each treatment. This is an important pre-requisite for statistical analysis. Without replications the estimation of error variance is impossible and there is no basis for statistical tests and for calculating confidence intervals. Also, replications permit a more precise estimate of the true mean response for the factors of the experiment. In addition, randomisations help to minimise the random variation of influence factors that are not investigated in the experiment.

Blocking is a technique used to improve the precision with which comparisons among the factors of interest are made. This is optional but can help to get to the bottom of what is really going on in a process. Here several experimental units are arranged in one block so that each treatment occurs once. Within each block there are relatively homogeneous experimental conditions. Often blocking is used to reduce the variability induced by factors that are not investigated in the experiment. Blocking implies that randomisation is not complete but restricted, i.e. each block is randomised separately. Blocking is, for example, useful, if there is a gradient of environmental factors in the experimental area, e.g. as introduced by a slope (Montgomery 2013).

Particularly in production ecology, medicine and technical sciences *controls* can be very useful. Here one treatment is included that actually is not a treatment but rather a "do nothing" option. In medical experiments, this is, for example, included by a placebo and in plant science the control may involve no management, e.g. applying no weeding or no fertilising. The control can help to answer the question what happens in the absence of manipulation, i.e. what happens, if the process is left to natural devices. Also treatment effects can then be expressed relative to the results of the control, i.e. how much more do we gain from applying a drug compared to a situation where no prescription is taken. Thus the control acts as a useful reference that can also be of benefit in the communication of research results.

It is a much debated feature of scientific experiments that they often tend to raise more questions than they answer and as a consequence hypotheses tend to get increasingly narrow as science advances. Reducing a phenomenon to a series of rather narrow hypotheses is a basic requirement of the traditional scientific method (Kimmins 2004).

Experimental design is a vast field of knowledge and research. A comprehensive treatment of the subject is beyond the scope of this book and the reader is referred to the literature, particularly to Montgomery (2013). Research students (MSc and PhD students) are recommended to take a course in experimental design and ANOVA (analysis of variance) right at the start of their programmes to ensure they develop at least a basic understanding of these complexities which can later be refined in statistical consultation. The design of the experiments and of the data collection planned in everybody's research should be fixed at the *beginning* of the project along with the procedures of subsequent statistical analysis. Otherwise one might learn that the data collected with considerable effort and expense do not fulfil the requirements of the statistical analysis, when it is too late. Along similar lines authors of scientific papers often make elementary errors in their statistical analysis. Research students should therefore take classes in statistics and consult statistical specialists that can help to avoid serious errors.

2.3   Modelling

An alternative to but also a complement of analysis is modelling. Here I refer to the narrow sense of the term 'model' involving mathematical models. The traditional scientific method may occasionally tend to encourage analytical reductionism at the expense of synthesis and integration (Kimmins 2004). Modelling synthesises existing knowledge, e.g. the knowledge gained from a scientific experiment, but can also venture into new directions, where little is known. It is, for example, possible to incorporate much of the complexity of an observed system into a comprehensive model and to study the model behaviour in order to make inferences about the real system. Some models are *empirical* or *statistical*, e.g. those that are directly based on the results of ANOVA, others are closer to the actual processes and these are often termed *mechanistic* models. Models complement analysis, because they can be used to reproduce the conclusions made in the analysis by simulation. Models are also applied to make predictions, e.g. about weather or climate change in the future (*forecasting models*) or about some process in the past (*backcasting models*). Alternatively, models may be built as a purely heuristic exercise, a way of exploring and synthesising what we think we know about some system and of identifying what we do not know but would like or need to know (*explanatory models*) (Kimmins 2004). They also play an important role in testing theories in ecology and physiology.

An important principle in modelling as well as in hypothesis testing is to 'keep it simple'. This requirement goes back William of Ockham who in the fourteenth century stated: *Pluralitas non est ponenda sine necessitate*, i.e. entities should not be multiplied unnecessarily (Gosling and Noordam 2011). This principle is also known as *Ockham's Razor*. Suppose, for example, you have two theories that predict the same thing then Ockham's Razor would discount the theory containing unnecessary information. This principle merely establishes priorities, it does not guarantee that the simpler theory will be correct. It also applies to modelling, where an important criterion for efficient models is *parameter parsimony*, i.e. not to unnecessarily increase the number of model terms and parameters. This principle is also part of modern statistical characteristics such as the *Akaike information criterion*.

3

Doctoral Studies and All That

Abstract The matters discussed in this chapter are particularly dedicated to those who are considering a PhD position and to current doctoral students. Having said that you will also find suggestions here which may be relevant to MSc students concerned with a research-master degree and to any other early career researcher, too. Guiding undergraduate and postgraduate students with an interest in progressing towards a PhD and doctoral students who just started by discussing some of the items in this chapter is an important part of academic tutoring and mentoring.

   This chapter may help you to decide whether research – either for earning a PhD title or as a long-term career – is something you seriously want to consider. I also share some thoughts about the choice of research topics, study location, making the first steps towards building up a research profile. This is followed by a discussion on research courses that are offered to doctoral students.

3.1   To Do or Not to Do: a PhD

Many students decide to stay on in academia after their BSc and MSc degrees and engage in doctoral studies. This decision is not easy and should not be taken on a whim. Sometimes they do this because their friends do it, sometimes they follow their parents' advice. Naturally there is a host of other reasons for engaging in a PhD. To help people make an informed choice about entering doctoral education a mentor who has doctoral experience can be beneficial

(Hopkins et al. 2020). Before making a final decision it is helpful to ask yourself the following questions:

- Do you really need and want a PhD?
- Are you a creative person with sufficient imagination and vision (or wish to become one)?
- Are you an idealist, a person who is keen on life-long learning and discoveries?

The first question is something you need to be able to answer with a definite "yes", if you want your doctoral studies to be a success. As with other things in life, you can only win, if you really want it. For example, a bad starting point is to accept the offer for a PhD position simply because you need a job, where any other job outside academia may have equally met your requirements. Working on a PhD is something special requiring more than the average attitude and commitment towards a job. It is a bit of a quest really. To be admitted to a PhD degree programme clearly is a privilege, far not everybody succeeds in securing a doctoral position. If you decide to take up the offer of a paid PhD position just for the income, both you and your supervisor will eventually be disappointed.

The question about pros and cons of doing a PhD, however, also relates to the objectives that you associate with your PhD, e.g. are you only interested in the title or are you aiming at a long-term research career as a kind of long-term quest. Whilst it is perfectly fine to follow up a PhD just for acquiring a title, this may potentially give you a little less motivation than a genuine interest in and fascination for the research topic. Naturally PhD titles can be useful for many jobs in private companies and in state administration and it may be your strategy to win the title just as another qualification. Again, this is totally fine. On the other hand, it is also absolutely legitimate to use the PhD time as an opportunity to find out what you really want to achieve in life and what your personal career path may be.

You will also need to be a bit of a pragmatic idealist, since you are frequently engaging in some "detective work" to get to the bottom of your research questions. You also need to be prepared to re-invent yourself several times over as part of your engagement in *life-long learning*, as continuous learning is an important part of everybody's research career. A scientist typically is open-minded and prepared to drop assumptions that they thought to be safe, if new knowledge suggests so. With many scientists this often becomes a life style.

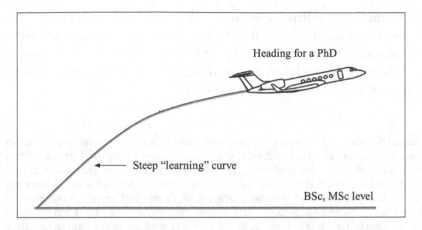

Heading for a PhD

Steep "learning" curve

BSc, MSc level

Fig. 3.1 The metaphor of a plane take off

Using an air traffic metaphor, the question whether someone is able to successfully finish a PhD shows similarities with the question whether a pilot can get his plane to take off from the runway (Fig. 3.1). Engaging in a PhD is fundamentally different from anything most students have ever done before in their undergraduate and postgraduate degree programmes. In a similar way, a novice pilot may have only driven cars before but never ventured into air space. A doctorate is unlike any other mode of academic study requiring *autonomy* and *self-direction*. Many beginners find this a somewhat scary experience. To some extent this also relates to students doing a *research master*, i.e. a non-taught MSc degree where the student is only concerned with working on their master thesis during the entire degree. This is comparable to a mini PhD or a licentiate degree in the Swedish university system and, incidentally, is a good preparation for a PhD. Your supervisor will be something like a personal tutor directing you towards the autonomy you are expected to develop. Supervisors are supposed to help supervisees to transition from a formal, taught, course-based programme mentality to a more independent research mentality (see Sect. 3.5). As part of this, supervisors need to make clear at the outset that their supervisees are responsible for their projects and that doctoral research is much about their taking on the responsibility of making an original contribution to the field (Hopkins et al. 2020).

Particularly the creative part of a PhD's research work, where one needs to develop one's own research questions and corresponding experiments, is often experienced as a difficult obstacle. For some developing this *creativity* is a steep "learning" curve. And precisely this part, the creativity part, is impossible to learn in courses (see below). That someone can meet or even embrace this creativity makes the difference between "taking off" and remaining at ground level. Once a research student has "taken off" in this way, s/he will succeed and finalise their thesis. It is not a shame at all, if someone is not able to "take off". This means s/he is better at something else and attempting a PhD is the ultimate way to find out what one is made of.

> Several decades ago not much support was given to doctoral students in pursuing their PhD. Originally the basic idea was that they were supposed to learn from the "master", i.e. their professor, by watching closely how s/he was carrying out their job. Incidentally, this has also been the traditional way of learning and training apprentices outside academia in China and Japan. There was no formal supervision and certainly no mentoring in those times. Doctoral students were expected to learn from watching others and by mimicking what they did. At the same time many doctoral students were "exploited" as personal (teaching) assistants. Despite the crudeness of this approach good students would eventually (often in fact after many years) "take off", others would not and be lost. This was perceived as a system that would select the better students who would then in time become promising researchers (Gosling and Noordam 2011, p. viii). Needless to say that this approach did not favour disadvantaged students that joined from other countries, had to work part-time outside the university or to take care of small children. Luckily this system has been abandoned over the years and several measures were put in place that would help PhD students on their way towards graduation (see Sect. 1.1).

It probably does not come as a surprise that following up a university career, will never make you rich. Salaries in private companies are usually much higher and jobs are more stable there. University careers can even turn out to be "stony" paths and disappointing at times to say the least, see Chap. 9. Despite your best efforts and although this is unethical, you may come across colleagues and superiors trying to prevent you from succeeding or deliberately holding you back. As in other workplaces, there may be even instances of bullying, though any university nowadays has services in place that support you in such unfortunate events.

However, the biggest asset of university careers is the unique opportunity of highly creative and inspiring jobs and of enjoying more freedom than in any other type of employment, even when your academic rank is still low. You also have the unique chance of pushing the boundaries of your understanding into unknown territory every day, i.e. you live a life at a virtual frontier. Long-term research careers also involve a lot of travelling and living abroad (at least for some time), as you will become highly specialised and the scientific job market

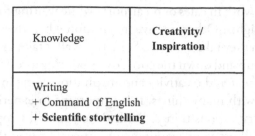

Fig. 3.2 Important research career components. Creative elements are set in bold

is international. Again, all or some of this may be appealing to some people and a headache for others. Those who are "made" for research usually love it.

As we touched on earlier, for a PhD and even more so for a long-term career in research and academic teaching you need to be creative and resourceful. Knowledge and the ability to learn and remember facts alone are not sufficient (see Fig. 3.2). The ability to work independently and creatively is also required. I have known many researchers with excellent in-depth knowledge of their subject area who failed to produce research papers and proposals, because they lacked creativity and inspiration. Both elements, knowledge and creativity, are very important and complement each other. Figure 3.2 highlights important elements of research careers that require creativity. Creativity is about devising new research questions, new theories, developing research plans and arriving at conclusions that advance the research field. Creativity can be encouraged by good mentors and requires an environment that gives researchers space, freedom and peace of mind. Unfortunately creativity cannot be taught, but it can grow. The Book of Taliesin by Gwyneth Lewis and Rowan Williams (2019) offers a good definition:

> Creativity/inspiration is better thought of as a *state of altered consciousness* in which the researcher receives knowledge of matters beyond what can be routinely learned (~ Lewis and Williams 2019).

Creativity often comes to us as a kind of intuition that almost feels like a faint prophesy from outside our mind. This manifestation of creativity cannot be forced and often appears in unexpected situations. Sometimes it just requires a certain environment, voices in the background, music, the smell from a cup of coffee nearby or others talking to you about seemingly unrelated matters. This is why some people have surges of inspiration when they once in a while happen to work in cafés or on airports while waiting for research guests or for their own flights. Others receive inspiration when they go for a walk in the woods or when they take a shower. Sometimes all it takes is rising from your chair and walking up and down the corridor or pouring a coffee in the canteen. Everybody is different and creativity and inspiration come in different shapes and disguises. As with many things, it is important to experiment with your creativity and to try various things. With time you will happen to recognise patterns and can adjust your working style in such a way that you increase incidents of creativity.

There is the famous anecdote of J. R. R. Tolkien writing the words "In a hole in the ground there lived a hobbit" that marked the beginning of the novel "The Hobbit" on a spare leaf of a student's examination paper. In an interview with Denys Gueroult in 1965, Tolkien remarked that he never hardly got through any fairy story before he felt the urge to write one himself. This is a remarkable statement and a strong sign of creativity. I experience very much the same every time I am reading other people's published work or carry out peer reviews. It is then that seemingly out of nowhere all sorts of research ideas come to me and I feel the strong desire to develop my own model or run my own analysis, see Sect. 5.3.

Many lucky people have inherited the gift of creativity and inspiration without doing anything to acquire this skill. Sometimes they even do not know that they possess this gift and discover it only by accident later in their lives. However, it is not true that you either have creativity or you do not have it. Creativity can be fostered and nurtured by parents, teachers and mentors. This is the most interesting but also the most challenging part of academic mentoring.

Quite frequently it can occur that you are stuck and your mind is blocked. Then it is definitely time for a break. Walking away from your computer for a short while can help to unlock your mind. Letting your task go for a moment and doing something completely different activates your subconscience and all of a sudden the solution comes to you on its own accord.

This observation often reminds me of the story of Pwyll Prince of Dyfed and Rhiannon in the First Branch of the Welsh Mabinogi (Davies 2007):

Pwyll Prince of Dyfed was lord of the seven Cantrefs of Dyfed and once upon a time he went to his chief court at Arberth. He was sitting on top of a mound, called Gorsedd Arberth, that was above the palace, when he saw a lady, on a pure white horse of large size, with garment of shining gold around her. The horse seemed to move at a slow and even pace. Pwyll decided to meet her, mounted his horse and followed her as fast as he could. But the greater he spurred his horse, the further she drew away from him, although her pace was no faster than before. He tried again and again and the more he urged his horse, the further was she from him. Since he saw that it was futile for him to pursue her, Pwyll said, "O maiden, for the sake of the man you love most, wait for me." "I will wait gladly," she said, "and it would have been better for your horse if you had asked me a while ago. I am Rhiannon, the daughter of Hyfaidd Hen."

The abridged story is a nice, canny metaphor for the fact that science as well as relationships cannot be rushed and takes time. Particularly creativity and inspiration must be allowed to enter our work on their own accord without pressure.

Creativity and inspiration are crucial to identifying new research fields and to devising research questions. Part of the process of doctoral research is becoming an *independent researcher* (Hopkins et al. 2020). Once you have become the leader of a research group, however small it may be, you are supposed to independently develop your own visions and research questions.

It can be argued that a research career is in fact a progression towards more and more (scientific) freedom and independence at work until you have become a senior researcher or even a professor.

This progression is only possible through creativity. Creativity and inspiration are also important in scientific storytelling which is required in research proposal and paper writing, see Chap. 5. Here, expected or incurred research outcomes need to be embedded in a pleasant narrative that attracts and entertains readers. Scientific storytelling also plays an important part in oral research talks, teaching and in research proposals.

Most people describe the time of their PhD work as the time of their lives: Never again you will have so much time to acquire new knowledge and skills, whilst being comparatively free in making your own decisions. Use this time wisely and above all: Enjoy it!

3.2 Research Topic

In contrast to earlier times, there is an increasing trend for the research topic to come with the PhD post and cannot be changed, because the prospective supervisor has secured funding for this specific topic. In that case you need to consider carefully, if this particular topic (and the corresponding position) is really your cup of tea. Particularly with regard to a possible long-term research career it is important to choose carefully, since this choice may largely define your professional future. Incidentally, do not apply for a wide range of different PhD topics at the same time just for the sake of securing a post regardless of the topic. The world is smaller than you think and such a strategy may cast serious doubt on your motivation. Choosing one's research field is never easy. The following questions may be helpful guidance:

- Do you really feel passionate about the research field proposed to you? Is this "your thing" or has somebody talked you into it?
- Is this research field/topic nationally and internationally widely acknowledged? How much recognition does the subject get, particularly by funding bodies in different countries?
- Will the topic help you to develop a long-term scientific career? Approximately how many research posts, lectureships or professorships are likely to be available at the time when you will apply for such faculty positions?

Since this is a crucial choice, it is better not to leave this entirely to chance or to personal relationships. Talk to as many people as possible about these questions. Involve peers and experienced researchers in these discussions but also your family. There is no simple way of giving "watertight" recommendations. Your choice should also reflect current research trends. Once you have made your choice, go for it and do not let anybody deter you.

Generally speaking it is a good idea to observe research trends already as an undergraduate and postgraduate student. Discuss them with peers, senior researchers and mentors. It cannot be emphasised enough that, to a large degree, the topic of your PhD determines the general research field you will be working in throughout your life, if you will continue to work in research. Therefore be careful with the selection of your specialisation, since some topics are clearly more favoured by long-term research trends than others. Such "favours" can involve more and better funding and a faster career progression among other things. However, even after a certain research field has been determined there are at least some small possibilities to steer your research into a direction which is favoured by current research trends.

There are always fluctuations in the popularity of research topics. Particularly the question whether *basic science* is more important to society than *applied science* has often been discussed and the discussion is still ongoing. Currently we live in a time where basic science is more supported, particularly by state funding bodies. Basic research related to climate change and human health are of particular interest to state funding bodies.

Excellence initiatives and excellence clusters, that have started to be launched 10–15 years ago, promote basic sciences and increasingly dislocate applied sciences that are then taken up by new *universities of applied sciences* and colleges (see end of Sect. 8.2) or by the industry. Governments and local universities support the excellence initiatives and try to free up resources from other subject areas.

Having said that freedom of research is part of national legislation in most countries, so you are naturally free to choose your own research topic. However, in the long run, it is hard to pursue a topic that is not much supported by national and international funding policies.

In recent years, when it comes to natural sciences, it has therefore been useful to select a topic that is at least related to basic science. It is a good idea to keep an open mind and also to engage in *interdisciplinary* research involving research communities outside your own field. On the other hand, it is also important that you really love the research topic and not behave purely opportunistically, since only true love and enthusiasm for a certain topic will provide enough motivation to carry you through difficult times, which are sure to come. Some researchers try to behave in an *anticyclic way* by anticipating important future research trends and topics, but this naturally involves a lot of guesswork and intuition not to mention associated risks.

Equally important as defining a general research topic or direction is the definition of detailed research objectives of the individual studies they are embarking on (Newton 2007). A failure to define aims with sufficient precision inevitably leads to poorly focused research, the lack of clear results and potentially a great deal of wasted effort. Newton (2007) provided a brief checklist that applies to both research topics and objectives:

- *Is the topic original, novel?* – Anybody worked on this question before? Check `Google`, `ISI Web of Science`, contact institutions.
- *Is it feasible?* – Can the question be answered given obvious restrictions. If not, does focusing or tightening the objectives help? Avoid "impossible missions".
- *Is it interesting, topical?* – Media interest, number of papers/conferences dedicated to the topic, availability of funding, policy relevance. How much long-term potential?
- *Can the objectives be phrased as a question?* – Helps in focusing the design of the research and in looking for answers. Sub-questions to break the problem down into manageable clearly defined units.
- *Is it of practical value?* – Sometimes important, but carefully check. Relationship to theory is more important.

Time spent refining objectives is never wasted. Seek advice from your peers, colleagues, supervisors and mentors before embarking on the project. Observe and analyse how the objectives are described in published scientific papers (Newton 2007).

3.3 Where to Go and Format of PhD Thesis

Obviously, aiming at a well-respected university for your PhD that is different from the universities where you have been awarded your BSc and/or MSc degree(s) is a good idea. Often it is, however, more common that you want to join the research group of a famous professor or other academic and not all of them are based at famous universities. Then you need to make up your mind, what is more important to you. Usually it is seen as a plus in academic job applications, if candidates made an effort to earn their degrees at different universities and in different countries. To facilitate this was an important consideration of the Bologna process, see Chap. 7. Therefore it is commonly advised to go abroad for some time of your career, particularly when your kids are still small, provided a good job is offered to you there. Times abroad usually make a positive impact on your cv. When you have grown up in a country with a national language other than English and then choose to work for an extended time in a country with English as the native language, this will inevitably foster your skills of expressing yourself in English orally and in writing. Such a longer stay in an Anglo-American country but also in countries and regions such as Malta and Hongkong will give you an important long-term advantage in your research career.

When for some reason you go to live in a country, where English is not the first language, learn the national language of that country as this makes the difference between being integrated in society or not. The knowledge of the local language opens many doors that otherwise remain closed. Nobody expects a perfect command of the national language, understanding and getting by is sufficient to show that you made an effort. If your kids are still young, they will also learn the local language quickly, bring friends home and may need help with homework in that language.

If you have a choice, prefer a cumulative PhD thesis to a traditional monograph. In that case you publish whilst working on your PhD, i.e. you engage in two important and required activities at the same time. Another benefit is that your thesis work has already been subjected to a detailed quality check (i.e. the reviews organised by the scientific journals) before it is submitted to the faculty and to external examiners. Also, the chances of unintended plagiarism are reduced, when you organise your work in a number of different papers. In my personal experience, doctoral students who have written a cumulative PhD thesis are better prepared for an academic career and are more experienced postdocs. Cumulative doctoral theses are almost a standard nowadays, particularly in natural and life sciences.

When doing a cumulative PhD thesis try to work on two papers at the same time (but on different days of the week) to ensure a constant flow of publications. In doing so, alternate publication projects that take a lot of time to finish with others that require less time.

Sometimes it is a good idea to get started in a new research field or topic by writing a literature review, see Sect. 5.3. Then you learn more about the breadth and depth of a subject, get to know the relevant literature and develop a better understanding for where your specific research questions fit in. Review papers are often cited much more frequently than research papers, i.e. they can rapidly increase your citation index, see Sect. 5.1.

Do not leave the submission of your research papers to journals until last. Start submitting your papers as early as possible to benefit from the experience of the review process so that your supervisor and mentor can guide you in the submission process if needed. Getting a paper successfully through a review process is a vital experience and skill that you need to learn along with carrying out the actual research and writing scientific texts. The continuous submission of papers also gives you more job satisfaction, since you feel that more and more of your "mission is accomplished". If some manuscripts are still unpublished after graduation, make sure you get them published soon afterwards. You really need them and cannot afford giving up on them after all the effort. The same often applies to the department or the research group you were based in during your PhD. They usually would like to have your work published, too. Then you can ask your supervisor and/or mentor to help you with funding during the time of finalising your last papers, perhaps your department can grant you an extension of a few months to your position.

The most difficult part of any cumulative doctoral thesis is the *synthesis*. This work goes by different names and is sometimes wrongly referred to as "summary". Usually the synthesis forms the first part of your thesis demonstrating how all of your papers are linked. Often students confuse the synthesis with a summary and literally write an abstract for each paper. That defeats the purpose of this exercise: The synthesis is meant to present your papers in the wider context of the associated research community. For this purpose you adopt a kind of "bird's-eye" view, characterising the wider research landscape at a higher level and critically showing how your results fit into this landscape.

Along similar lines as the synthesis, in your *PhD defence* you are supposed to give an interesting, challenging research talk quite similar to a conference presentation. Again do not present summaries of your papers and particularly do not structure your talk according to the order of manuscripts or when they were published, e.g. "In paper 1, I carried out…. Moving on to paper 2, I studied …". Nothing could be more wrong and boring. Obviously you need to discuss this with your supervisor and mentor, however, it is much more meaningful and entertaining to focus on the big picture, the wider research context and present highlights of your research. It is also a good idea to include a short personal story about how one day you at last hit the nail on the head and found the missing link that really opened your eyes for the drivers of a particular process you were studying. Definitely secure the support of your supervisor and/or another experienced researcher including your mentor for preparing you for your PhD defence. It is important to practise your final talk several times before delivering the final speech (see Chap. 6).

3.4   How to Build up Your Research Field

I have already stated that a research career in a way is a journey towards increasing freedom and independence at work, see the end of Sect. 3.1. As a PhD student and young researcher you learn very quickly that depending on others for contributions can slow you down. For example, relying on someone else to carry out your lab analyses or for programming simulation code can take considerable time, since the person you rely on may be very busy and you end up in a long queue with others. Sometimes this is unavoidable, but better try not to run into such situations. Instead developing skills that allow you to *bypass bottlenecks* puts you into a situation where hardly anything or anybody can hold you back. Being able to bypass bottlenecks is a crucial resilience skill.

Later when you move from junior to senior researcher and eventually become professor, each increase in rank will come with more responsibilities and challenges but also with more freedom. This freedom is essential for carrying out ground-breaking creative work. Therefore, as stated in Sect. 3.1, moving up the ranks is not a journey towards more power and particularly not towards lots of money, but a journey towards independence of others who can interfere with tasks and directions and a senior researcher clearly has the responsibility to put this independence to the greatest possible use for the benefit of academia and society.

Often you come into a situation where you set tasks for yourself or you discuss tasks with your line manager or supervisor. Here it is important to observe that all objectives and corresponding tasks satisfy three important conditions (Speidel 1972; Gadow and Bredenkamp 1992; Gosling and Noordam 2011):

- **Measurable**: The description of the objectives/tasks need to be clear an unambiguous so that they can be measured,
- **Attainable**: Objectives/tasks must be realistic, they must be possible to attain,
- **Time-related**: Objectives/tasks must have a clear time frame with deadlines for their attainment.

In other words, there is no point in setting a goal that you cannot measure, cannot attain or is not realistic. In practice, there are often violations against these principles. Assessing the success of an activity is difficult if not impossible in such a situation. It is good practice to check all objectives that you set for yourself, all tasks that are given to you (e.g. in performance reviews) and anything you ask others to carry out against these criteria.

Often you can even compensate for a lack of supervision by turning to colleagues and respected scholars of your networks or to other members of the supervisory committee. This way you are in a position to ensure that even in the worst-case scenario, when things go wrong or turn out badly, you will succeed anyway. Reassure yourself that there is always a way forward and around problems. Ask yourself the questions: "What can I do to turn the tide despite the odds? What can I do myself to make this research or publication happen anyway?" This way you build up valuable resilience and resourcefulness for use in difficult situations.

If your research happens to involve programming and creating experimental software, do not spend excessive time in designing nice user interfaces, help files etc. This is, of course, fun, but it also takes much of your valuable time. Always remember that the scientific question and the final publication(s) should be at the centre of your attention. Software like R (http://cran.r-project.org/), NetLogo (http://ccl.northwestern.edu/netlogo/) and/or Shiny (http://shiny.rstudio.com/) allow you to develop sophisticated simulation code without much effort. A larger analysis application or modelling software is only worth the effort provided it will enable you to publish at least 5–10 significant papers. This has always been my "rule of thumb" and it has worked well for me. If your scientific software later turns out to be a major breakthrough and selling point, hire a private company for re-designing the code and for software engineering. Sometimes you find that your institution is keen to help with funding such efforts that contribute to research dissemination and outreach.

Another strategic question relates to long-term (environmental) monitoring. Some departments or research groups run long-term monitoring programmes with repeated re-measurements. Such long-term data can be very valuable and interesting for research. Funding is not always easy to secure for long-term monitoring. Similar issues relate to the long-term maintenance of a scientific model/simulator or a large data base. These activities require a lot of valuable time and also involve work on aspects of minor scientific importance thus potentially decreasing your publication outputs. Current research policies with project lifetimes of 3–5 years do not greatly support this kind of activities, although they certainly have many merits and benefits. Some research groups spend an excessive amount of time on collecting data and convert only a tiny share of them to scientific publications. Remember that it is also possible to cooperate with people concerned with long-term data collection and support them in writing papers on their data or model. As always it is crucial to strike a healthy balance between input and output. Do not engage in research work just for the sake of it, carefully check the potential outputs.

Some research fields of natural and life sciences are related to a certain industry branch. Carrying out applied research that is taken up by the industry is naturally very rewarding and motivating. Although you may then be concerned with applied research, try not to allow the question of research methods to be determined by how practical they may be for use in practice and industry. First, the most important consideration is to select the scientifically best method for the general problem at hand (see Fig. 3.3). Once the problem is scientifically sufficiently explored, it is always possible to identify ways of simplification later or to produce a kind of "data adapter" to facilitate the uptake of your findings by practitioners.

As mentioned before, in some research groups, a specialised scientific software or a monitoring system are maintained that need to cover a whole region or country. In such situations it is not uncommon that "research topics" are proposed with the sole purpose of solving a technical deficiency of the system under consideration. This, however, does not constitute a valid research question unless the technical problem is generalised and the resulting general research question is separated from the technical problem. The technical issue can be addressed as part of research dissemination (see Fig. 3.3).

If you need to acquire new knowledge for research or teaching, the traditional way to do this is to consult papers and textbooks. Wikipedia or online teaching programmes including YouTube may offer fast access to the topic so that you can better see the big picture.

As part of designing your own research field build up national and international networks. Such networks will be invaluable in your research career and always stay with you wherever you go. With the people you engage nationally and internationally you will write research proposals and publications together. They may help you with future job applications or act as mentors. It also feels good to have colleagues elsewhere in the world you can cooperate with, when for some reason cooperation temporarily does not work out so well at your home campus at a given time. A good way of starting networks is to secure

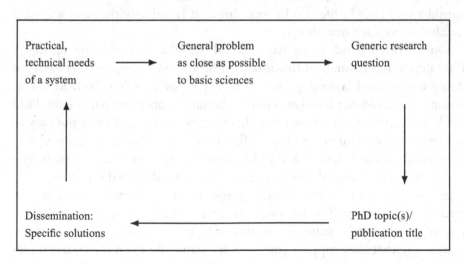

Fig. 3.3 How to turn a practical or technical "research" problem into a real research question

a place in a PhD programme where you have supervisors from two different countries and need to move between them. These programmes are usually well-funded and provide successful applicants with a stipend for travelling, computers and books. Start collecting future collaborators by staying in contact with the students of your MSc classes and with those that you meet during your PhD studies.

If this is not anyway part of your PhD study plan, try to include conference and teaching experience. Future employers want to see some evidence of this experience. Discuss this with your supervisor. However, do not engage in too much teaching or other activities (e.g. committees, consultation etc.), as this may distract you from your research work. If possible sign up for an assessed (pedagogic) teaching course in higher education and obtain an official certificate. This will qualify you for future faculty jobs that you may want to apply for after your postdoc time.

3.5   Research Courses

When engaging in a PhD degree you are usually required to collect course credits. Here you need to balance strategic credit collection against the need to learn new skills and to gain knowledge. Strategic behaviour, for example, involves the question: How do I get the necessary amount of credits as fast as possible with as little effort as possible? Here, of course, you also need to consider that later in life and in your career, it is unlikely that you will have much time to learn new things.

On the other hand, many students fall into the trap of believing that a PhD degree is a natural continuation of BSc and MSc degrees: Just carry on doing courses and collecting credits, then you are fine (see Sect. 3.1). This assumption could not be more wrong. The most important part of any PhD work is independent, creative research and paper writing. This is not easy to accomplish when you do this for the first time. Therefore it is good advice to get on with this task as soon as possible. Doing PhD courses is only partially a way of filling educational gaps. Summer schools are designed for filling larger gaps. Also, successful course participation or finalising many courses does not guarantee that you will be successful in carrying out your research. A lack of research work/papers cannot be substituted by course credits. Select courses in such a way that they support your research work. Be open for course topics that are very new to you, even if you are a little afraid of them. International research courses (e.g. those funded by NOVA in Scandinavian countries or by the EU) usually operate at a high level and are particularly worth pursuing.

They are excellent for networking and are also great fun for organisers and lecturers.

A senior, well accomplished colleague with a solid record of successful PhD students once told me over a pint of beer:

> Sometimes I supervised PhD students who came to me with distinctions in their BSc and MSc degrees. They always had the top marks in their classes. So they thought they would master a PhD as well. When they eventually got going, they knocked at my office door every second day to ask what would be the next step in their research work. They wanted me to spell out every single task they had to do. Supervision turned out to be a difficult process with these students.
>
> Then I had other students. With these I discussed possible research questions in initial meetings that I usually have with every doctoral student. The students left my office after the meeting had concluded and never returned for weeks. I already started to suspect they had given up on their PhD studies, when one by one they returned with glowing eyes and greeted me with the words: "Professor! I took you up on that research question you mentioned in our last meeting and I have really found something!" Many of these students did not even have good marks in their previous studies, but it was easy to supervise them and to get them to finish their doctoral studies successfully.

When realising the need for acquiring additional knowledge or new skills (e.g. in statistics or in another subject area you have not studied so far), there are six main options:

1. Self-study,
2. Courses/summer schools,
3. Consultation,
4. Publishing with a senior author,
5. Attending specialised conferences/workshops,
6. Visits to/internships in other research groups.

Picking up books and studying the new subject matter yourself is certainly a good option and what you have learned while reading these texts will stick for a long time. However, not every student can do this in isolation. Some need discussions, activities or simply the contact with other students and lecturers. Everybody is different.

Courses including those given in summer schools are a very good option to gain broad, fundamental knowledge in a new field. Sometimes, however, you are pushed for time or have a certain research plan in mind. In that case a course may be too broad and may take too long. Then a point effort would provide more focussed help like as part of specialised *consultation*, e.g. in statistics.

Consultation involves seeing an expert or specialised person that you can ask detailed questions about a field that is new to you and where you got a little stuck in your research. However, it is not the aim of consultation that consultants do a certain task of your research for you. Consultation just offers specialised advice and acts as *catalyst*.

Ask around and find out, whether your institution offers consultation in various fields of science, e.g. in statistics, modelling or proposal writing, and what is implied.

You can also learn by doing and team up with a senior researcher in the new field you are interested in. The senior partner will then guide and teach you whilst you carry out the experiment, do the calculations and write the main text. This guidance and teaching is, of course, informal and materialises in discussions about the manuscript and/or in email comments and answers. This is a very good way of learning, however, it does not suit everyone. From my personal experience I can say that this is an option I have often chosen with great success, particularly when, as a young faculty member and parent, I could not afford the luxury of frequently going away for an extended visit or attending a specialised course.

Some researchers attend specialised conferences or workshops for learning new skills, e.g. they attend a workshop in Bayesian statistics applied to freshwater ecology, in order to understand how this field of statistics may help to better analyse data or carry out modelling in freshwater ecology. If you follow this strategy, you can talk to presenters and extend your professional networks to include specialists from the new field.

Another option is to visit other research groups that have expertise in a field new to you that you would like to know more about. If you can afford to stay longer, e.g. by means of a dedicated travelling grant or stipend, you can even turn this into a kind of internship. As part of such an internship it is, of course, possible to pursue mutual publications with members of the host organisation whilst having the benefit to talk to the experienced hosts face to face.

All these six options are certainly not necessarily successful in isolation. Often you need to apply a combination of them depending on the situation in hand and on your personality.

4

Early Career Years

Abstract Early career years are a period of continued qualification between your PhD degree and your first permanent faculty position. What you learn during this time typically complements and extends the skills and knowledge you acquired in your PhD studies. This is quality time that will never come again and you should use it wisely whilst your responsibilities are still fairly limited. What you pick up in this time may be crucial to any senior position you will later apply for. A postdoc period usually lasts for 2–3 years and it is good practice to go abroad for that time or at least to change university. Special mobility grants usually help to achieve this.

During the postdoc time you continue laying the foundations for your professional future. Although you are still young and just have completed your PhD, this is a very important time in your life that usually coincides with founding your family.

4.1 Collecting Experience and Records

Start looking for a postdoc position 1½ years before the end of your PhD. It does not matter, if some of your papers have not yet been published by that time. Speak to your mentor/supervisor and ask them to help you. Do not leave this matter to chance, be proactive. When the funds have been exhausted at one place, you can also carry out another postdoc period elsewhere, if you feel you need more time or no permanent position is available. In that case make sure the two postdoc periods are based at different organisations and you thus have the chance to learn many new skills.

Postdoc periods are meant to consolidate the knowledge and skills you acquired during your PhD work. This period is extremely valuable and comparable to the tradition of *journeyman years* dating back to medieval times in Europe. They were a time of travel for several years after completing apprenticeship as a craftsman. In a similar way, the idea is that you get exposed to the work and thinking of different academic schools. At the same time you disseminate the methods, skills and ideas you acquired as an MSc and PhD student to the new place. This academic exchange of ideas and skills is essential for academia to thrive. During the postdoc time you can mainly focus on research. Never again in your career you will have so much time for research. Towards the end of your early career years you should move on to your first permanent academic/faculty position.

This is an important period in your career when you start to develop your own *research vision* and *strategy*. You not only assume more responsible roles in the institution you work for, but also start to see more clearly which path you possibly want to take. At the same time you can compare the career path and personality of your PhD supervisor with that of your new line manager at the place where you are spending your postdoc time. You are now part of a different school of thought and also have the opportunity to speak to different peers. All this gives you ideas and makes you think. The information you gather puts you into a position to coin your own research vision and strategy, which you will need when applying for a faculty position as lecturer, researcher or research group leader and attending job interviews.

During your postdoc time regularly check out university websites and subscriptions like `science-jobs-de`, `academics.de`, university jobs in sciences (https://www.academickeys.com/) or `jobRxiv` (https://jobrxiv.org) once a week. There are also great offers on `Twitter` and on `Facebook`. It is a good habit to stay informed about current developments on the academic job market, so that you are prepared should the need for a move arise. Always have your personal records, e.g. your cv etc., up to date so that you can quickly respond to any attractive job offers and do not lose valuable research time for writing job applications. Setting, for example, 1–2 hours aside for updating your personal documents every six months is a good idea.

For your cv, it is good to collect certificates from different universities with excellent reputations. Therefore do not always stay at the same place, keep moving. Having done this myself several times I know this is easier said than done, since after a number of years you form a kind of attachment to a certain place and the requirements of your partner and children are also to be considered. In academia, changing workplaces is now perceived as an essential requirement demonstrating a person's flexibility and potential.

If possible and English is not your first language, go abroad, particularly to an English speaking country to improve your command of the language of science. Having been abroad is usually viewed positively by home universities you want to apply for permanent jobs later in your career. Before deciding where to go, check out life and institutions in that country and what future job opportunities are there for you and your partner.

Write or speak to well-known academics in your area and ask whether they have or plan for a new postdoc position you can apply for. Sometimes it helps to introduce yourself personally. Conferences, workshops and summer schools are also a great opportunity to discuss upcoming jobs with potential project leaders. Offer to help with proposal writing for your postdoc position, if there is no upcoming job, but a potential new supervisor is willing to take you on provided you will receive a grant.

In the worst case accept work experience *placements* or *internships* as a starting point. You can then later try to apply for an open position from "within the organisation". While you are working as an intern, you are starting to make valuable experience and get to know people that may help you in the future.

It pays off to check what are the requirements for the next career step. You may need to attend certain courses, get involved in PhD assessment/supervisory committees, write another dissertation, get involved as external examiner in evaluating degree programmes of other universities among other things. Such requirements typically differ from place to place. Although many of these activities sadly distract us from the things we are really interested in, particularly from research, they may be necessary to make you eligible for advancing your career, securing permanent employment and/or gaining more academic freedom.

Particularly when you are starting in your career, review research manuscripts for scientific journals and get involved in the board of a scientific journal. This is also a good way to stay up to date in your research field and to learn new skills of scientific management. While reviewing manuscripts you pick up information on new findings or new methods. Decline review requests from scientific journals or funding bodies only for very good reasons. After all you learn from these activities how to write good paper and proposals by reading other people's work. At the same time it is good to get involved with scientific journals as a reviewer, since you may want to submit your own manuscripts to them and then it can be helpful that the editors know you.

Invite distinguished colleagues from abroad to your institution and get involved in research exchange programmes, e.g. COST[1] actions. Also organise conferences and workshops or assist in organising them. This is a good way of extending your networks and of getting known in your research community.

Always be keen to acquire new skills and learn about new subject areas by writing publications together with senior specialists in those areas and/or by preparing a new teaching module on these topics. When there is a need to write about a new subject or to teach it, you will inevitably learn yourself what it is about. Get also involved in professional organisations and societies.

Keep detailed track records of your annual achievements including publications, (invited!) conference participations, taught modules, invited lectures, reviews (journals, books, proposals), journal/committee work and other activities. This helps you in discussions with your superiors, e.g. in annual performance reviews or as part of forward job plans, and allows you to do a self-reflection on what you have achieved.

4.2　Leadership

Leadership does not come easy to most people. Yet some are longing for it. Others have a talent for leadership. However, researchers do not typically follow a career path that fosters leadership skills. Few courses are offered on this subject and academics are usually not well prepared for leadership roles. Many books have been written on this subject and here I just draw your attention to a few important points that have always served me well in my career.

It is a good idea to keep a record of mental notes of all the things you liked and disliked in other people's leadership, when you were a postdoc or young faculty staff. Be observant all the time and most importantly include yourself in these observations. Whenever you think a certain leadership behaviour, on balance, was not right or you were in the mood of complaining, make a note. Also, of course, when you experienced good leadership, try to remember this. You can learn from both positive and negative experience and this is something to come back to when developing your own leadership style so that you avoid the negative bits when it is eventually your turn to lead. This "record keeping" enables you to overcome whatever seemingly bad happened to you or others

---

[1] The European Cooperation in Science and Technology (COST) runs an EU-funded programme enabling researchers and innovators to set up their own research networks in a wide range of scientific fields.

by better actions on your side. It will keep you busy and prevents that you to get obsessed with the injustice you experienced.

When you eventually come to power, lead mostly by example and extend kindness to everybody irrespective of rank, do not lead by office. Usually those are the best leaders in academia who do not long for power, but just do the job, because it has fallen to them and there was nobody else at the time to take on this responsibility. People must follow you because they like you, because they trust in your leadership and most importantly because you inspire them with your academic accomplishments. You must earn the loyalty of your staff, not demand it. Thoroughly prepare for this situation whilst you are not in power. As a leader the thought "What can I do for others put into my care?" should be constantly on your mind.

At all times remember that a university is not an army: Leading by command is not possible and contradicts true academic spirit. You depend on the good will of others and you need to inspire and to care for them. Any attempt to lead by force is likely to backfire. Along similar lines, never try to make your staff work on something they do not like or are not able to do. Recognise their strengths and give them tasks according to these strengths, even if you originally hired them for different tasks. Above all give your co-workers sincere credit for what they do well.

5

Scientific Storytelling

Abstract Publishing research results unites scientists across all disciplines. Writing of scientific texts is the foremost task of any researcher and we typically spend approximately 90% of our time composing texts for a wide range of different purposes. Most of the time we devote to this crucial task is about presenting our thoughts and results in the clearest way possible whilst forging an intriguing story at the same time. Here much creativity is required and when writing you need to create an atmosphere that allows creativity to blossom.

The ability to write good texts does not come easy to most of us and it is something we need to work on throughout our entire career. In addition there are a few technical issues, rules and strategies to consider when writing research papers that not everybody is familiar with when embarking on a research career. Scientific storytelling is also required when writing research proposals.

5.1 Composing Scientific Texts

People working in academia are doomed to write many texts every day: Books, reports, research papers, research proposals, popular research notes, press releases, policy documents, reference letters, lecture notes, examination papers, job applications and job adverts, module description and syllabus documents not to mention the flood of forms that need to be filled out and also require short texts. This implies that every academic needs to be a writer of sorts, a word smith with a particular skill to forge texts that are spot on for a wide variety of different purposes. Often these texts have to be written in different languages, at least in English and in many countries also in another, national language. This requires great skills in creative writing and in linguistics. Therefore many doctoral schools and programmes at universities have responded to this requirement by offering courses in scientific writing as part of the courses offered to research students. This has been a good move, since we really spend much more time on writing than we do on actual research work.

Communication is an important part of professional and academic expertise. To put it blankly, research is of limited use unless the results can be passed on for others to assimilate, develop or implement (Price 1998). You can have the best ideas in the world, devise good plans and make excellent scientific discoveries, but if you cannot communicate them clearly enough to be published, your work will not be acknowledged by the research community (British Ecological Society 2015). Through the communication in appropriate scientific journals and publishing houses, we and the organisations we work for receive official recognition. Still, it is true that most of our scientific work passes by unnoticed by the majority of other researchers in our research community let alone by those from other research communities. However, if we keep publishing "through official channels", i.e. in good scientific journals and publishing houses, there is at least a fair chance that some of our research work will be appreciated by others and cited. If we do not publish, our research work has no chance at all to be noticed.

When communicating orally or by writing texts, researchers must clearly express their ideas. To do this effectively and depending on the purpose of communication they need to develop different *storylines* for the research they are concerned with or they are planning to do. Storyline is the way a researcher makes sense of their project for their audience. Others call it a thread that runs through a project and brings coherence to all aspects of the research project presented. Importantly, storylines change according to the audience and according to what you hope to gain from telling the story (Hopkins et al. 2020). Naturally this also applies to teaching (Chap. 7) and scientific presentations (Chap. 6).

Without doubt the most important writing that academics are concerned with are publications. Both an individual's and the university's scientific strength are measured mostly in scientific publications. It was not long ago that in some research fields publications were something researchers concerned themselves with only, if there was time left after fulfilling other responsibilities. Only a few years ago an older colleague even told me that publications are something we only do for prettying up our personal cv's, but they would not contribute to the greater good of a department. Occasionally you still hear such opinions, but they are hopelessly outdated. On the contrary, every research group, department and university need as many high-quality papers in different scientific fields, as they can possibly get. Local and national reviews of research activities mainly focus on research papers. Sometimes universities even encourage researchers to publish by re-distributing funds according to publication activities. Often researchers are prepared to put a great share of their spare time into producing publications and universities and similar institutions "exploit" the self-interest of each individual. To some degree the institutions we work for honour this by the promotion of individuals and by granting them freedoms in their work. Publications are also documents that will always stay with you no matter what happens. As much as they are important to your institution, they also form a crucial part of your cv. Often you need to leave teaching modules, experiments, research plots etc. behind when you move on to another place, but nobody can take your publications away from you.

It is very important to stress that publishing is both a fundamental right and responsibility of any researcher. Let no one deprive you of this right and fulfil your publication duties as best as you can. By publishing you are achieving three key goals for yourself and the larger scientific endeavour, i.e. *disseminating your research, advancing your career* and *advancing science* (British Ecological Society 2015).

Try to consider publishing and following characteristics such as impact factors and citation indices etc. like a sport or like following the stock exchange. Adopt the habit of loving it. Only if we love something, we will truly be successful. At the same time set yourself a target of at least 2–4 good English-language publications in ISI[1] journals per year. If possible, the number of publications should be evenly distributed over the years. If you have the choice, rather publish research articles in scientific journals than in book chapters.

In November 2005, J. E. Hirsch published the h index, a famous citation index, with the objective to characterise the scientific output of researchers (Hirsch 2005). h

(continued)

---

[1]International Scientific Indexing.

denotes the number of publications of a given researcher that have been cited at least h times and all other publications have less than h citations. If a researcher has an h index of 20, all of her $h = 20$ publications have been cited ≥ 20 times, whilst all other publications of this person have < 20 citations. To progress to a higher h index becomes harder and harder with every increase (Ritchie 2020).

The h index can be calculated based on bibliographical data bases such as Scopus, Web of Science and Google Scholar and is not recommended for early-career researchers. As many other citation indices also the h index has a number of weaknesses. The index, for example, much depends on the citation culture in different scientific fields and should not be used to compare researchers that belong to different research communities. Self-citations can influence the h index and the number of co-authors and respective contributions as well as the author rank are not considered in the calculation. A scientist's h index is often explicitly taken into account for hiring and promotion decisions (Ritchie 2020).

In natural and life sciences, books (monographs) are less valued than scientific papers published in good journals. These are the main avenue for publishing research results (British Ecological Society 2015). This is likely to change now with ongoing pandemics that require more distance learning and textbooks to be written. In any event it is advisable for any researcher to write at least one textbook in their career. This is a laudable effort contributing to important discussions in the research community, synthesising one's work, contributing to teaching and helping young researchers to build their own careers. It is also a valuable documentation of your scientific school of thought and of your research group. Writing a textbook is a valuable experience, since you pull individual research results from different papers together and in the process understand how they all fit together and where there are gaps. If you make the effort, it is important to engage in this task long before retirement (ideally when you are between 40 and 50 years of age) so that the book is actively used in teaching and research while you are still working at your institution. This will also help to write new, refined editions. Each textbook should ideally have several editions and not just one. Textbooks usually are prepared by many years of research and publications. They are partly based on your and other people's peer-reviewed articles and partly on your teaching material. This way it is ensured that the majority of contents of a textbook is peer-reviewed.

Sometimes you hear that writing papers can only start after securing funds for a project and after the first data of this project have been collected.

This is not entirely true. In many cases writing good papers is possible even without research funding, e.g. by writing a review or conceptual paper or by collaborating with other researchers. On the contrary, some of the best papers have not received any public funding. Therefore do not worry, if you do not happen to have research projects for a while, keep applying and most importantly keep writing papers. There is always something that can be done.

If you have secured funding for a larger project, package the outcomes in separate, small papers dealing only with one research question at a time. This way you do not overload each manuscript and you also ensure that the project produces a sufficient number of high-quality papers. Lengthy manuscripts involving too many research questions are in fact not well received by editors and reviewers. On the other hand, packaging your achievements into too many small papers, is also not professional (Ritchie 2020). Though you should publish several papers every year, it is worth keeping in mind that it is not so much the number of publications but more their quality and the corresponding journal reputation which really matter. Make it a rule that, whatever else happens around you, you always assign the highest priority to publishing.

When planning a manuscript aim to invite at least one co-author from whom you can learn both in terms of the subject matter at hand and in terms of language. Cooperating with authors from different countries is fun and sometimes it is easier to work with them than with some of your colleagues on your own campus because of a healthy emotional distance. Some of these co-authors can help with the English language, others have good, creative ideas and yet again others may help you to develop a better understanding of the subject matter or of statistics. Cooperating with authors from abroad also has the added benefit that you potentially can include a data set contributed by one of your co-authors. This may help to avoid getting the label of a "local study", which is sometimes assigned by reviewers to manuscripts that report on national data only. Also, at some universities, co-authors based at the same institution (e.g. the same department) need to share the credits of their mutual publication in university performance reviews. However, if all or most of your co-authors are from outside your university, most local credits naturally come to you. This is not always important and obviously you want and need to publish together with your team and work colleagues for a number of good reasons, but for some performance exercises this is worth keeping in mind.

In devising your research questions, keep a healthy balance between re-searching fundamental problems on one hand and researching method devel-opment and applied matters on the other. In the long run the citation indices of your own work can support your decision making, i.e. they tell you how your work was received by the international research community.

Occasional publishing in good journals outside your subject area increases your interdisciplinary visibility and improves your career chances, particularly if your original research community is rather small. Avoid terms specific to narrow subject areas (or at least define them) and jargon used by representatives of your field when submitting manuscripts to such journals.

5.2   Authorship

According to international rules a co-author of a scientific article can only be who substantially contributes both to the analysis and to the text. Others who helped in different ways or provided only data should be mentioned in the Acknowledgements. General supervision, participation solely in the acquisition of funding or the collection of data strictly speaking do not justify authorship. Of course, exceptions can be made, if particular circumstances apply. Also refer to the Vancouver protocol (https://research.ntu.edu.sg/rieo/Documents/Foundational%20Documents/Vancouver%20Protocol.pdf)   for orientation. In principle all co-authors should be able to give a spontaneous presentation of the research questions, methods and major findings of a recent paper if requested (Binkley, pers. comm.).

Some suggest negotiating the order of authors and their roles early on and even to fix agreements in formal documents, see the discussion on *prenuptial agreements* (*science prenups*) in the internet. This is, of course, possible and makes sense where there has been a history of disagreements involving authorships in the past, but in my personal experience it has never been necessary. With many journals it is now customary to describe the author roles (authorship contribution statement). This is good practice enforcing transparency and helps to reflect on the input everybody in the team of authors has provided.

You sometimes read or even hear that some people or organisations only make their data available for analysis or modelling, if they are considered as co-authors in the forthcoming paper without any intention to take part in the data analysis and/or the manuscript writing. However, in most countries around the world, research data in general but particularly data that have been collected using tax payers' money are legally public and should therefore be made available on request. In the UK, this is, for example, guaranteed by the *Freedom of Information Act* (FOIA). This is why you should not accept conditions attached to public data such as the inclusion of authors who have contributed to the data collection but have not been involved in the analysis/paper writing, see Chap. 10. Make such people aware of the fact that their claim is illegitimate and then move on by approaching other people. Some organisations attempt to bypass the freedom of information legislation by arguing that they incur costs when making the requested data available and that they are therefore either not in a position to meet the request or need to charge for preparing and releasing the data. This is strictly speaking not acceptable either, see Chap. 10.

Generally, author order is determined by the level of contribution of each author (Hopkins et al. 2020). The first (and usually also the corresponding) author should be who contributes at least 70–80% of the total work. The last author is usually the head of institute/head of research group, a senior scientist or mentor/professor provided s/he has substantially contributed to the paper. The ranks in the middle are often perceived as minor ranks in the author list. In some disciplines and when there are many authors, alphabetical order rather than contribution order is followed. In some countries such as China, the role of corresponding author also carries an important weight.

In some institutions, it is traditional practice that the head of institution or head of research group is always included as co-author, even if s/he has not contributed to the publication in any way. Note that this is not acceptable practice any more.

Increasingly you see papers with many authors. Generally it is a good policy not to aim at including too many co-authors. After all, how is it possible to coordinate and practically write a consistent paper with more than five authors? Also it is hard to justify author contributions and roles in that case. A maximum of 3–4 authors in total is a good guideline. The less the better, since this also puts more weight on your own contribution and subsequent institutional credits.

It is possible (but not always advisable) to form "strategic publication alliances" with partners in different universities/countries so that you alternate the sequence of co-authors in order to share the workload whilst ensuring a constant flow of papers. This can be a good solution, particularly, when teaching loads and other responsibilities are considerable. Just make sure that you do not "collect" too many authors and that this strategy does not attract "freeloaders" who never contribute much. While you are not so well-known as author why not team up with a senior author of international reputation? At the same time you can learn from her or him.

Finally, for performance reviews and for building your own career, it is also important to be first author on at least half of your publications. First-author papers also grant you the opportunity to drive your own agenda and to live your own vision. There is nothing wrong with joining a group of authors that are working on a publication, however, from time to time you simply need to be the lead author.

5.3   Processing Literature

Searching for and reading scientific literature related to a particular research question is an important part of our research work. Staying on top of things in one's research field also means a lot of reading. Ideally you try to keep reading all the time. Most importantly you do not want to risk doing work that someone else has already published. Increasingly, publishing houses and various online services offer publication alerts. Colleagues in your networks and in your research project team may point out important papers to you and there are also regular alerts on Twitter and Facebook. Another way of staying up to date with publications is, of course, to be involved in journal work as reviewer and editor.

It is a good thing saving time that you hardly need to physically go to the library any more but can access the relevant journals in the internet or carry out electronic searches on whatever you are interested in. Even when you are travelling or working abroad you can easily connect with your institutional VPN (Virtual Private Network) and download whatever you need provided you have internet access. In electronic searches, you need to be creative with the key words that you type in. If one set of words does not give you good results, devise and try another one. Do not give up or finish your search too early. The citation index can advise you on how often a certain paper has been cited. This can be a good indicator of important papers (Gosling and Noordam 2011). It is good practice to optimise your reference list by citing and referring to important authors, seminal papers and textbooks and to papers in high ranking journals. For fundamental concepts and definitions cite textbooks rather than papers. Make sure that most of your references are recent to demonstrate both that you have a good understanding of current literature and that your research is relevant (British Ecological Society 2015).

A regular *literature seminar* in your research group where one or two colleagues present a published study that was carried out by authors outside your own research group, may also help to keep everybody up to date and is very motivating at the same time. Literature seminars support team building and can also be an interesting teaching tool, see Chap. 7.

I personally find it helpful to check out papers and books first for the big picture, i.e. the research topic, the theories and the outcomes, to gain a quick overview. This does not take long and most of this you understand by reading the Abstract and parts of the Method section. Then I usually take a break and afterwards decide whether it is worth reading the rest of the publication. Often it is sufficient to read just the bits and pieces you need for making progress in your own research or for inclusion in your manuscript.

> Why not keep a *pdf archive* where you collect all publications you read and find useful? Store their titles in an appropriate front end such as `EndNote` or `Mendeley`, to enable you to do quick searches. You can also link these data bases with the one of your supervisor or with colleagues in your networks. In collaboration with others it often happens that you need to share the pdf files of articles with others and having a well-organised archive makes this task easier and faster to accomplish. You can also share your literature collection and associated front ends with the students you teach. Have your pdf literature archive on your computer, a USB stick, on an external hard drive or in a cloud so that these resources are with you all the time and on every device.

When reading electronic copies or hardcopies of your papers always keep your computer in reach when reading a paper: Make digital notes in a computer file, not notes on paper, so that you can copy/paste them for use in your own paper and thus save time. If your university library has not subscribed to a particular journal, ask around in your networks, if someone can access and send you the paper you are looking for. Also ask your supervisor. Incidentally, reading literature is not only reading for contents: Read scientific papers written by native English speakers also for the way how the paper was structured and which phrases, syntax, punctuation and words were used. Try to remember elegant and useful phrases for possible use in your own papers.

Often reading of other people's papers sets your imagination free: All sorts of research ideas and research questions come to your mind while reading related literature. This particularly often happens to me when reading manuscripts as part of a journal review process or student examination papers. Make notes of these ideas, they are precious and later consider whether they are worth following up, see Section 3.1.

A *literature review* can be a small part of a larger research paper that you are writing, but it can also be a free-standing paper in its own right. Some journals or part of journals have specialised on publishing literature reviews only. Although literature reviews do not constitute research in the narrow sense of the word, they usually collect many citations, as they are useful starting points and references for other authors, so make a habit of writing a review paper from time to time. A literature review may simply be a factual report on what has already been written on a topic. In this case your job is to summarise and collate the information on the subject. To do so requires a degree of generic skills in information processing, i.e. how to extract the main points and how to structure related sets of information. It also requires sufficient knowledge of a broad subject area to understand the essential meaning of what has been written.

Particularly in free-standing reviews you may be required to supply an evaluation of what has been written. In this case you have to weigh the merits of apparently contradictory results of statements, give reasons why some of these deserve more attention than others and possibly offer a personal view about where the truth, on balance, appears to lie. Clearly to do this well needs a deeper knowledge of the subject, probably with some first-hand experience in this field (Price 1998).

There are three basic ways to structure a review. In the first, the literature is dealt with author-by-author (at worst in alphabetical order, better in historical succession) and paper-by-paper (possibly in date order). Such an approach is likely to be safe and comprehensive, but perhaps a little uninspiring. It is most likely to be appropriate for short descriptions of the state of the art in the Introduction of research papers, where readers just want to know what has happened so far and possibly make up their own minds.

The second approach, appropriate to an evaluation of the literature, goes through the subject area sub-topic by sub-topic, preferably in a logical hierarchy. The material is arranged in such a way as to highlight possible conflicts, ambiguities and consensuses in the literature. These are commented on with perhaps a suggested resolution (Price 1998).

As a third approach Price (1998) suggested a hybrid which is more interesting than the first approach, but less prone to charges of author bias than the second. The material is introduced sub-topic by sub-topic and by author in historical succession under the sub-headings.

In terms of organising your work, clearly the first phase is to read or at least to scan the literature. There are various ways of identifying suitable literature:

1. Look up the reference lists in recent books and articles,
2. Simply google suitable topic descriptors and key words,
3. Use more specialised science engines,
4. Ask your mentors and supervisors, if they can add to your list of books and articles.

Start reading with the Abstract, this often gives you a quick idea, whether the paper in question should be included in the review. You will probably discover that there is a relatively small number of key papers which are referred to by many other authors or which seem to make most of the relevant points. As well as information, they may well supply you with appropriate sub-topic headings (Price 1998).

Structure your review as you go along: If any new paper that you find requires a modification to your structure, simply do it. It is better to start writing right from the beginning, as you feel a sense of progress and any word processing software makes it easy to introduce changes to the structure of your text. At some point you need to draw a line under your data-gathering efforts, even though it is likely that you have not included all relevant literature.

A more formal way of carrying out literature reviews has been suggested by the research community concerned with "*evidence-based methods*" (e.g. Sutherland et al. 2004; Livoreil et al. 2017). The basic idea of this approach to reviews is to conduct a systematic, quantitative appraisal of the evidence provided in the scientific literature through meta-analysis techniques. As part of these techniques, searches are transparent and reproducible and thus minimise biases (Livoreil et al. 2017; see also the resources on http://synthesistraining. github.io). The results of one reviewed paper can be thought of as a single data point in the meta-analysis. These techniques also involve strict criteria for assessing the quality of the data in each study – a process called *critical appraisal*. Also, a hierarchy of evidence is commonly used in evidence-based research, where the findings of studies using strict experimental designs are given greater weight than those having no comparison or control elements (Sutherland et al. 2004). Such systematic reviews will probably be the standard for literature reviews before long.

5.4   Getting Started

When starting with a new research subject, either as a PhD student or whenever there is a need to move into a different field, it may be a good idea to write a literature review as your first paper to learn what your research area is actually about and what kind of research has been done already in this field, see Sect. 5.3. This will allow you to see the whole field, to understand the state of the art and to identify gaps that you may be able to fill.

Let us assume that you have already clearly defined your own coherent research strategy and vision and advertised these on your website and in job interviews, as is good practice. Now quite a few research colleagues know you for this research field and associated vision. Try to make all your papers be related to this strategy of yours so that they contribute new knowledge to your field. Naturally you need to explore new directions every now and again and boundaries between research fields are not that strict any longer, but it makes a somewhat confusing impression, if the topics of your papers are all over the place with no clear red thread.

Starting a new publication project is a psychological challenge for many people. To ease this process, always start with a short *research plan*. This will also help you to reflect on how the forthcoming new research relates to your overall research vision and strategy. Research questions should prominently feature in this plan. You can start developing first ideas on the white board in your office, perhaps together with your team of co-authors. Take photos of these white-board ideas as they unfold and later transfer a refined version to a word processing software where you can also develop a rough structure, into which ideas and existing sentences can be filled and modified as inspiration comes to you. Then develop the Methods and Results sections. Abstract, Introduction and Discussion are the last sections to write. They require a lot of care, are closely related to one another and often final details are only clear towards the end of paper writing after you have interpreted the results. The research plan also allows you to see more clearly where your co-authors come in with their expertise. Share the plan with them. Then everybody knows what they have to do and also what the deadlines are.

When you are faced with a large piece of work, it is easy to become daunted by the immensity of the task (Price 1998). As with many things in life, the trick is to break this immensity down into little steps or parcels that are manageable. This is where the aforementioned research plan can help. In psychology, this technique is referred to as *compartmentalisation*. Compartmentalisation is hands-on stress management reducing anxiety and tension and everybody can do it.

> Compartmentalisation is a psychological defence mechanism to avoid mental discomfort, stress and anxiety. There are many types and meanings of compartmentalisation and here I refer to the ability to divide up our tasks, responsibilities and thoughts so that they do not overlap and compete for your attention all the time.

A most basic form of compartmentalisation is, for example, to compile a "to-do list" and to strike through each task that you have accomplished one after another. Even the simple compilation of such a list contributes to sorting your thoughts and to calming you down. Carnegie (1998) drew an analogy to safety technology that allows ship crews to immediately shut off various parts of an ocean liner into watertight compartments to avoid total flooding in the case of an accident. In a similar way compartmentalisation enables us to shut off the past, future or any other conflicting thoughts that may deter us from our current task. Instead he suggested to live in "day-tight" compartments by concentrating with all our intelligence and enthusiasm on doing today's work superbly today.

As part of practical compartmentalisation focus on only one thing at a time and do not allow to be distracted by thoughts about other tasks during that time. It may also happen that a comment from a work colleague or an email reminds you of something uncomfortable in the past. Compartmentalisation then means that you do not dwell on this memory or on its trigger but rather let it pass. Another recommendation is to focus only on what you can control. For example, the fact that the printer that you urgently need to use broke down, is out of your control. However, you can control your response and try to see something positive in the technical failure: This gives you an opportunity to check what other printers nearby you can use instead. You always wanted to check this anyway. Turning the event into something positive reduces stress. It is also part of compartmentalisation to switch off unnecessary distractions such as visual and sound alerts on your computer, mobile and other devices. When you finally are at home after work, do not allow work related thoughts to impact on the focus on your spare time, simply enjoy it without mental distractions. So with compartmentalisation in mind, do not sit there and reflect on the immensity of the work but get started with the first step!

Why not prepare a flowchart or graphical sketch of your research/experimental plan including your specific research vision and the research questions? Keep updating it while reading and writing. Later this flowchart may come handy in your paper as one of the figures or as graphical abstract and it will for sure fuel your imagination and creativity while your paper takes shape. Pin a copy of this flowchart/sketch on a wall where you can see it every day and update it as you go along.

Some researchers find it helpful to write short reports about their research progress, not more than 1–2 pages every second or third day. This is particularly helpful when starting your research career, but may also be handy even later. The reports will help to break down large research projects into small steps, which makes it easier to cope with the workload and with venturing into unknown research territory. They also make you understand where you currently stand, i.e. what you have achieved so far and what there still is to do. While writing you keep reflecting on your work and often new ideas enter your mind. You can also share the reports with your co-authors, supervisors and mentors. If English is not your first language, write these reports in English anyway, even if all authors share the same first language as you. The final paper will most likely also be in English and starting to describe what you do in your reports in English is a great asset and preparation for the final text. File these reports and also the response of your co-authors and colleagues along with drafts of your manuscript and important computer code snippets. This paper trail will prove to be a valuable resource and documentation later in your career whenever you need to go back to what you did then. Among other things it will also serve as an encouragement and reminder that you are able to take on and master complex research tasks despite the odds.

Incidentally, some guidance for getting started in individual-based and agent-based modelling has been proposed by Grimm et al. (2014). The authors also made useful suggestions for documenting the modelling steps in a way similar to the reporting discussed in the previous paragraph and referred to their guidelines as "TRACE" (TRAnsparent Comprehensive Ecological modelling documentation).

Before you begin to think about writing your journal article and where to submit it, it is important to thoroughly understand your own research and know the key conclusion you want to communicate. Consider the possible conclusion and ask yourself, is it

- new and interesting?
- contributing to a hot topic?
- providing solutions to difficult questions?

If you can answer 'yes' to all three questions, you have a good foundation message for a paper. Shape the whole narrative of your paper around this message and start with a skeletal plan around that narrative (British Ecological Society 2015).

Also, always write down spontaneous ideas in files so that you do not forget about them ("*brain dump*"). The more you advance in your research career the more such ideas come to you at unexpected times, see Sect. 3.1. Carry a small notebook with you at all times or simply log down these ideas in your mobile phone or tablet computer, but never let them pass unrecorded. A good way to store ideas is the *Zettelkasten* method, see https://zettelkasten.de. Although we so much draw on creativity in research, an excess of new ideas can sometimes also become a curse: Do not change your original research plan too often and rather keep some of the new ideas for follow-up papers.

I found it also useful to alternate manuscript writing and analysis for the same publication, because the ongoing analysis tells us what direction to take in the writing and likewise the writing can prompt us to try another avenue of analysis.

A colleague of mine recently commented:

"It seems crazy, but my experience has been the more novel my ideas, the more likely they are to get rejected from publications and the only 2 or 3 papers I've ever just sailed through the publication process were probably my least interesting papers. I have mentioned this to a few people recently and their responses have been: Me too."

This comment highlights one of the faults of our current peer-review system, i.e. that it sometimes fails to encourage *innovative research*, although this is exactly what we all are supposed to pursue. Despite this do not be afraid of coming up with seemingly "crazy" ideas from time to time, as such unorthodox thoughts may turn out to be really innovative and contribute to seminal papers later on. Have the confidence to go for them, even if some colleagues laugh or shake their heads. The little story in the previous box also shows that we are not alone with some of the problems we are facing and that it makes good sense to discuss them with others in our networks. Remember that sadly the value of many famous pieces of art and poetry were only fully appreciated long after the artists or authors had died. Whilst we all hope that some appreciation of what we do will actually come earlier, we are well advised to have confidence and to be patient.

What Bach means to me? Consolation. He gives me the confidence that in art as in life a true accomplishment cannot be ignored or suppressed, needs no human assistance but asserts itself through its own power, when the time has come. This confidence we need for our work, he had it.

Albert Schweitzer on Johann Sebastian Bach

To encourage and foster writing some institutions offer *writing retreats* or *'Shut up and Write!'* events. These events normally span several days and the purpose is to create a space, both physical and mental, for researchers at any stage of career to focus exclusively on writing. They are common practice in many research institutions and are enormously valuable for boosting your writing productivity. They may be an institution-wide event or organised at the school or department level and they can be held on campus, but also at an off-site venue, such as a country house, to ensure a tranquil environment (Hopkins et al. 2020). The mentors participating in such an event help each attendee to produce clear writing targets and writing specialists can be consulted on specific issues. Workshops covering writing-related themes can also be part of the retreat. Joining such an event may be a good option, if you get stuck in the writing process.

5.5   Structuring your Paper

Once you have devised a good research question or even a set of research questions think of an attractive *title* that may inspire or even provoke reviewers/readers a little. A good title will help you get citations and may even be picked up by the press (British Ecological Society 2015). For example, it is possible to turn the main research question into the title of your paper, e.g. "Do bark beetle outbreaks amplify or dampen future bark beetle disturbances in Central Europe?". The British Ecological Society (2015) maintained that titles phrased as questions get downloaded more but cited less. Why not write down a couple of title candidates in your research plan and share them with your co-authors? The title needs to be informative and interesting to make it stand out to reviewers and subsequently to readers. Titles can be finalised shortly before submission and can be modified during the review process. Even provocative questions can very effectively be used as titles. Avoid boring titles such as "On the question of assessing volume …" or "A contribution to the analysis of …". Do not be too modest when choosing a title, but rather be self-confident when devising the title of your paper. Gosling and Noordam (2011) distinguished between *dynamic* and *static* titles, where dynamic titles can include a key result of your research in a full sentence, e.g. "Traits mediate a tradeoff in seedling growth response to light and conspecific density" as opposed to "The role of traits in seedling growth response to light and conspecific density". Naturally the style of title also depends a little on your research community and you have to take the limits into account that are set by journals and corresponding authors and readers. Among other things the title plays a crucial role in determining a paper's impact. Further advice on designing titles is given by the British Ecological Society (2015):

- Write in statement form. When scanning papers, most people skip to the last sentence of the abstract to look for the key message, so consider making that sentence your title,
- Keep it around 15 words. Any longer or shorter comes with more chance of being rejected at peer review,
- Use important key words in the title that are different from the official key words you specified (remember key words are used by readers to discover your paper),
- Use punctuation to split the main message and qualifier/subtitle,
- Keep it general: Readers prefer titles that emphasise broader conceptual or comparative issues and these titles fare better both pre- and post-publication than papers with organism-specific titles,
- Do not use abbreviations even if they are familiar in your field. You should keep a broad audience in mind.

Written communication is like a bridge between writer and reader: Like a bridge its performance and efficiency depend on its materials and structure. A good structure enables a powerful message to be conveyed with minimum use of words (Price 1998).

For detailed guidelines on how to structure your paper you obviously need to carefully consult the guidance available for every scientific journal, as these author guidelines usually set out the article structure. Often you do not know the final outlet when you start writing your manuscript. Then go for a structure that is a kind of compromise of the structure of all the journals that will be potential outlets for your manuscript or simply use the structure of similar papers you or others have written in the past. Such standard structure will help organise your material in a logical order (Gosling and Noordam 2011). In a similar way guidance about how to cite literature and how to prepare the section "References" or "Literature cited" are provided. Here it is also useful to refer to a recently published paper in the journal selected as outlet so that you can format the references and citations in the text correctly.

The overall structure of a piece of written work has blocks or sections of more or less homogeneous material arranged in order. By breaking down your manuscript into smaller sections, you will be communicating your message in a more digestible form. Subheadings should guide the reader through your narrative. Write each subheading in statement form and use key words in the headings to increase the search engine optimisation of your paper (British Ecological Society 2015). The subject matter usually develops in a certain direction. Often the first section describes the problem, the second collates the facts, the third discusses deductions from and implications of the facts and the last draws conclusions about relationships and/or about desirable courses of action (Price 1998).

Between and within sections, it is good practice to work from the general to the specific, from large-scale to small, from early to recent. Also, it is helpful to adopt a sequence of causality, e.g. to move from physical factors (geology to climate to soil) through biological factors (flora and fauna) to human factors (historical, demographic, socio-economic and political). In a well-written piece, the sections and sub-sections are evident by the cohesiveness of the material composing them. Nevertheless, it is often worthwhile to use sub-headings to divide the material. They help you to check that your organisation of material is clear-cut, give readers an impression of how the material is arranged and break up text into more manageable chunks.

Paragraphs should also be used as a structural element to keep together all closely related points. Check that your paragraphs do not contain a wide miscellany of information. This defeats the whole point of paragraphing, while a small number of large paragraphs makes difficult reading (Price 1998).

Some research community have developed their own standards for reporting research in their field that are widely accepted and apply across a range of journals. It pays off to be aware of such standards and protocols. For example in individual-based ecology, Grimm et al. (2010) have proposed the ODD protocol (Overview, Design concepts, Details) for reporting individual-based and agent-based models. The protocol applies to parts of the Materials and Methods Section, but can also be presented in an Appendix or as Supplementary Materials.

Many authors start writing the Introduction of their papers, although it is also possible and sometimes even better just to make a few notes for the Introduction and then to proceed with the Materials and Methods section. For the Introduction the following advice may be helpful:

Recommendations for the Introduction section

1. To set the scene and to motivate why this research is necessary (1–2 paragraphs),
2. Motivate and inspire the reader (1 paragraph),
3. Present the state of the art (3–4 paragraphs),
4. Give your objectives/ hypotheses (1 paragraph),
5. In total no more than 2–3 manuscript pages A4.

The Introduction is challenging to write, as it needs to attract readers and to cover the background/state of the art at the same time. This can only be achieved by writing in a concise way that is also attractive. Here you also motivate your research, i.e. what made you carry out this type of work, what caught your interest (Gosling and Noordam 2011). At the end of the Introduction all hypotheses and objectives need to be clearly stated in a separate paragraph. When a senior editor assesses your paper for peer review, they will be looking at whether your question is one that is worth asking. In your Introduction, state that your research is timely and important and why (British Ecological Society 2015). Sample other well-written papers to get an idea of what are good introductions.

Typically the Result section of your manuscript only presents your results, whilst you interpret the results in the Discussion. You certainly explore all interesting avenues of analysis, but only include the most interesting and meaningful results. In the Discussion, you reflect on the Introduction, answer your research questions and refer back to the hypotheses and objectives stated at the end of your Introduction. You should also refer to similar work in this field and explain how your results confirm or challenge previous results (Gosling and Noordam 2011). In doing this you usually refer to some of your figures here and also to previous papers. Finally, the Discussion should broaden out to the general topic. Depending on the type of research paper and the journal you submit to there can be exceptions from these rules. The Discussion/Conclusions section should finish with an intriguing and inspiring sentence, but can also suggest ideas for future research. The same can be recommended for the Abstract. Discussion/Conclusions, Introduction and Abstract need to be linked and any statements made have to correspond to each other.

An Abstract is not a list of contents in free prose: It is a self-contained piece of writing which says, with great terseness, the same thing as the text. It is often treated as a separate piece of work, and may very well be published separately in abstracting journals, conference proceedings or in a data base (Price 1998). Abstracts are freely available and affect how discoverable your article is in search engines (British Ecological Society 2015). The Abstract will almost certainly be read by more people than the full text and those who read the full text may do so, because the Abstract catches their attention. Therefore after the title the Abstract is the most read part of your paper. Editors and reviewers read the Abstract before deciding to review your paper. It is worth taking great care with it (Price 1998). Do not use your Abstract to talk about anything that is not in your paper and do not cram it with details, it is not a mini version of your paper. The last sentence of your Abstract should communicate your key message. The British Ecological Society (2015) recommended writing your Abstract after you have written your paper, when you are fully aware of the narrative of your paper. The Abstract should articulate your new and interesting key message, outline the methods and results, contextualise the work and highlight how your research contributes to the field and its future implications. If no guidelines are provided by the journal you plan to submit to, structure your Abstract as follows:

Abstract template

- Sentences 1–2 set the context and need for the work,
- Sentences 3–4 indicate the approach and methods used,
- The next 2–3 sentences outline the main results,
- The last sentence identifies the wider implications and relevance to management or policy.

The final Abstract point is the most important of all in maximising the impact of the paper. It should synthesise the paper's key messages and should be generic, seminal and accessible to non-specialists (adapted from the Journal of Applied Ecology website https://besjournals.onlinelibrary.wiley.com/hub/journal/13652664/about/author-guidelines). Also be aware that some journals limit the number of words in abstracts (approximately 200 words) and that compliance is sometimes checked on submission when you are asked to paste your Abstract into a text field of the web-based manuscript management software. In this context, it is also worth pointing out again that key words should be chosen wisely. They should not include anything that is already part of the title and should enable internet search machines to suggest your paper for a wide range of search terms.

Be lavish with credits in the Acknowledgement section of your article. There are always people who have contributed to the paper with data, in administrative terms, with software or have opened your eyes in discussions during the coffee break. Do not hold back here, give those people credit and mention them. They deserve it and by stating their names you spread so much trust and happiness. You should acknowledge any help received in compiling your written work: Those who helped with experimental work, those who advised you on statistical analysis, those who guided you (your supervisor, mentor), supplied ideas, commented on earlier drafts and those provided financial support. In dissertations and books, it is also common to thank those that made your life congenial while the work was done and those who helped with production. All this is common courtesy. It is also mandatory in submitting dissertations to acknowledge any academic help received (Price 1998).

Appendices or Supplementary Materials should contain material which is relevant to the topic of the document, but which

- is so voluminous that it would obstruct reading the text,
- provides detailed examples of a general principle or process discussed in the text,
- has been taken from other people's work,
- covers related material which is not needed to follow the text to a conclusion.

It is tempting to shovel everything collected but not needed in the text into Appendices, which thus become a sort of verbal cold store. Resist the temptation to do this. Ask yourself whether any reader might want the information to support reading the document. If so, arrange it carefully. Appendices need not lead in any sequence from one to the next, but the content of each Appendix should be cohesive. The main text of your work should refer the reader to each appendix at the points where the Appendix is most helpful. If there is no such reference, the Appendix is probably best left out (Price 1998).

5.6 About Text and Style

Style is hard to define and even harder to teach. Some people have it naturally, others do not and have to acquire it. Style is a matter of appropriateness rather than correctness. One cannot say 'that is wrong style'. Nor can anybody tell you the secret of an excellent style. However, certain hints can be given about style in general and academic style in particular. A comprehensive style is a function of conciseness, careful structure, meticulous logic and accuracy in use of language. It is also aided by good sentence structure or syntax (Price 1998).

Academic style tends to be impersonal and you need to avoid expressions of opinion if possible. In particular, avoid dogmatic statements. Controversial assertions should be supported by evidence from your own studies (before the statement), from a reputable published source (cited after the statement) or from a detailed argument (following the statement). This applies even more, when you write about a field where you are a beginner and not an expert.

> **Avoid any of the following:**
> - Racy, journalistic style,
> - Colloqial words and expressions,
> - Strong expressions such as 'extremely' and 'very',
> - Humour, jokes,
> - Rhetorical devices such as 'Surely ...'.

If experimental results indicate a conclusion only weakly or if only the majority of writers in the literature favour a certain viewpoint, it is safer to use *hedging terms* like 'it may be the case ...' or 'it could be tentatively concluded ...' or 'may be suggestive of a possible relationship' rather than committing yourself (Price 1998).

Having said that you should try to interest the reader in what you have to say. This encourages attention and a favourable response. Interest is maintained by a good *structure*, which leads the reader on towards an identifiable destination. It is aided by *comprehensibility*, which maintains the momentum of reading and *conciseness*, which brings the reader to the conclusion with minimum delay. Interest is also generated by variety. Varying the length of sentences is a useful device. A series of short sentences can become irritating. A series of long sentences, in which subsidiary clauses pile up one on top of another, not only becomes rather difficult to follow but also may have the tendency to send the reader to sleep. A short sentence after a long one has immediate impact. The occasional use of metaphors may also catch the reader's interest.

Attempt to use a variety of synonyms where they exist and where scientific accuracy does not require one word and only one word. 'Maybe', 'possibly' and 'perhaps' are everyday examples. 'Nevertheless', 'nonetheless', 'on the other hand' and 'by contrast' are substitutes for the overworked 'however'; 'thus', 'hence' and 'consequently' for 'therefore'. Use a thesaurus to suggest alternative words when one crops up too frequently. On the other hand, consistency of usage is important in conveying an impression of well-ordered work. Where more than one convention of spelling, punctuation or referencing may be used, choose one and stick to it (Price 1998).

Tense of verbs should be consistent and appropriate. For example, details of experimental procedure should be treated in a tense according to whether it is a description of a generally applied technique or something which you have done, e.g.

> In nursery experiments, trees *are planted* in weed-free soil.
>
> but
>
> In this experiment, trees *were planted* into weedy soil.

Several good textbooks have been published recently on how to do academic writing (essays, theses, papers etc.) using the English language. When you are a beginner and particularly when English is not your first language, it is a good idea to purchase at least one of these books and use it as a reference. There is no need to work through all exercises suggested in the book, just scan the contents and read selectively according to your needs.

Never write a manuscript first in your national language and then translate to English. Always try to think in English right from the beginning no matter how bad you believe your language skills are. With every language comes a particular mindset and a different way of thinking that has to be reflected by your text. Incidentally, it often happens to authors with surnames that do not sound English that they receive standard recommendations from reviewers to improve the quality of English in their manuscripts, even if their English is perfect or near perfect. This is one of the flaws of the peer-review system. Just learn to live with such remarks as gracefully as you can.

> In the past, there was a strict rule to describe methods and procedures in the text using passive verb forms only, e.g. "The trees were measured twice, once at the beginning and once at the end of the survey period.". This rule has been relaxed, so that it is possible and often even better to use active verb forms instead, i.e. "We measured the trees twice, once at the beginning and once at the end of the survey period."

There are many different ways to present your results and reading many papers with objectives similar to yours will give you much inspiration for your own manuscript. Whatever you can express as a graph (as part of a figure), do present as a graph rather than as text or a table. Not only does this keep your word count down but a well-designed figure can communicate a key message more effectively than a text or a table (British Ecological Society 2014).

Tables take much longer to read, to explain and to understand than graphs. Using graphs is a way of data visualisation. Always devise clear and precise graphs. Software like R are very helpful for producing high-quality graphs. Both tables and figures must be able to stand alone and must be accompanied by an explanatory caption that enables them to be understood without the need to read the body of the paper. This information should not be repeated in the main text. The main text, however, should refer to the information in the tables and figures (Gosling and Noordam 2011). Usually table captions are printed above tables whilst figure captions are set at the bottom of figures.

At the beginning of your career you may want to ask native English speakers to take a look at your texts and to make corrections, if English is not your first language. At some universities there are also services available that help correcting English texts. Use them and every time, when native speakers correct your written English, try to make mental notes and to remember the patterns of your mistakes so that every time round you improve your writing style and eventually do not need these services any more.

Always cite and refer to English language literature where possible. If you need to refer to other language sources, give additional translations of titles in the references (in square brackets following the original titles) to make their use easier for other researchers who do not know that language, e.g.

Johann, K., 1982.   Der „A-Wert" – ein objektiver Parameter zur Bestimmung der Freistellungsstärke von Zentralbäumen. [The "A-thinning index" – an objective parameter for the determination of release intensity of frame trees.] Tagungsbericht der Jahrestagung 1982 der Sektion Ertragskunde im Deutschen Verband Forstlicher Forschungsanstalten in Weibersbrunn, pp. 146–158.

Always aim at a style which renders your paper to be as clear possible. Pause after finishing a sentence or a paragraph and go back and reflect on it: Is it really clear what I am trying to say here? Even to a novice or a non-specialist? – At the same time we have to write our papers as concisely as possibly. This does not have to be a contradiction. We need to adopt a style of wording that leaves as little doubt or room for interpretation as possible while using words parsimoniously. It needs time to get used to this kind of writing, but reading many papers written by others and practice will get you there. Conciseness is a very practical virtue: What you write in your working life will be read by busy managers and administrators or by scientists who have many other papers to read. Their time is precious and should not be wasted. It is desirable to cultivate the habit of writing concisely right from the start, but it is always possible to rewrite at lesser length what you have already written. Conciseness can be achieved at every level: If the whole text is structured logically, it is less necessary to make cross-references or repeat information. Within sentences some words may be redundant or length may be reduced by restructuring the sentence. Long sentences are not readily held in the mind all at once and hard to understand. Therefore 20 words have been suggested as a normal upper limit. If circumstances require longer sentences, they need to be broken up by strong punctuation marks. Long paragraphs and sections are psychologically daunting to readers, who subconsciously are often looking for 'milestones' to give them a sense of progress (Price 1998). Obviously one should also strike a good balance between clarity and conciseness. It is easy to condense into unclarity. Hotaling (2020) published useful guidelines for concise writing that are worth checking out.

During the writing process always be mindful of *plagiarism* and *self-plagiarism*. Within the body of the text, you must cite another researcher whenever you refer to his or her results, conclusions, methods or statements in your paper (Gosling and Noordam 2011). For correct referencing and citation details, check published papers and the guide to authors provided by each journal. It is very easy to commit these offences for a simple lack of carefulness: After a while you may not remember that you have taken one sentence from this paper and another from that book over there on your shelf, so better stick the references into the main text immediately. You can then compile the reference list later so that this task does not distract you from the creative writing process. Plagiarism detection software can easily spot your oversights. Journal manuscript websites usually include plagiarism software and manuscripts are now regularly checked for plagiarism as a matter of routine. Therefore, but also for general ethical reasons, make sure that you always give due credit to other people's work and even to your own. Add references to other authors immediately when you use their thoughts so that you do not forget. Also it is good practice to paraphrase the sentences of others so that they better fit into the context of your paper. Paraphrasing also conveys a more professional impression as opposed to copying the sentences written by other authors word for word. Still do not forget to refer to the source, even if you paraphrase.

All of us often tend to fall into the trap of omitting some steps in a thought process or an argument that are particularly obvious to us. As a consequence important connecting sentences may be in our heads but never make it on paper. Such explanations may not be obvious to quite a few of our readers. It is important to write them down so that everybody can follow our texts.

Also, do not jump too much from one thought to another, pay attention to connecting sentences so that your text is easy to follow and is fun to read. This can easily happen when you "convert" an oral presentation given at a conference or one of your lectures to a paper. The presentation's bullet points are then often turned into short, mostly unrelated paragraphs with no connecting phrases. Remember that even scientific readers expect a nice story to be told and get tired after a while when your text is too technical and boring.

Along similar lines do not take common terms and concepts from your subject area for granted. Increasingly subject boundaries, particularly in journals, become more fluent and research fields tend to merge. As a result many readers may not be familiar with terms known only to a fairly small research community or to a few specialists. If you do use rare or new terms, explain and define them or refer to other papers/textbooks where they are properly explained in English. Also, do not abbreviate terms too much in your papers. Some authors enjoy doing this, allegedly for the sake of saving repetitions and space. However, as you may have experienced yourself, it is a pain to read a paper with an overkill of abbreviated terms where you need to go back to the beginning of the paper and look up the meaning of abbreviations in nearly every paragraph. This is bad style and to some degree obscures the contents of a scientific article. If you feel you need to use abbreviations, try not to come up with new ones (if avoidable), but rather use established ones that have a chance to be familiar to quite a few readers.

5.7   Logic

Logic is central to good academic and professional writing. In this section, I largely follow the excellent description by Price (1998), a former, much respected colleague of mine at Bangor University (Wales, UK).

Statements are always *facts*, *premises* or *deductions*. A fact is something scientifically incontrovertible (so far), as that the vegetation period is shorter at Umeå (Northern Sweden) than at Bangor (North Wales). In scientific writing, you may need to quote evidence or a written source. A premise is an assertion that your readers can be expected to agree with, e.g. 'All else being equal, mitigating climate change is desirable'. A valid deduction follows inescapably and without exception from premises and facts, e.g. 'Therefore it is desirable to plant more trees near Bangor rather than near Umeå.' This sentence has the form of a deduction and may well be true, but it is not a *valid* deduction. Further facts such as 'tree growth and survival increases with length of vegetation period' and premises such as 'planting trees is the best way of mitigating climate change' are needed before the deduction can be supported.

Also, check whether combinations of words in your sentences are used logically. Words such as 'and', 'moreover' and 'similarly' connect ideas which support each other. Words such as 'but', 'however', 'although', 'nevertheless', 'yet' and 'on the other hand' connect ideas which are in opposition. Sentences like 'Numerous studies have simulated climate-change scenarios.' are logically not correct, as studies are not in a position to simulate anything.

Common logical errors are listed below. Some of them have Latin names reflecting their recognition in Classical times Price (1998).

1. *Non sequitur* (= it does not follow)
   In a sense this covers all logical errors. Although the statement usually concerns the same subject matter as preceding statements, there is in fact no necessary connection. The deduction about tree-planting near Bangor, for example, given above is *non sequitur*, as is 'People are starving in the world, therefore we should subsidise Swedish agriculture' – a logical fallacy often thought and not seldom spoken.

2. *Ignoratio elenchi* (= ignoring the point)
   This entails proving what is not disputed or disapproving what no-one is asserting. It represents an attempt to draw attention away from the real subject of debate by holding up an irrelevant conclusion as though it is an affirmation or refutation of what the argument is really about. Thus, in a discussion about reducing carbon dioxide emissions by travelling less and carrying out more video-conference meetings someone might argue that a university wastes more money on lights and computers that are not turned off in university buildings after work.

3. *Ad hominem (or ad feminam)* (= to the man (or to the woman))
   This involves an attempt to discredit the person making an argument rather than the argument s/he is making. For example, 'pay no attention to the lecturer who argues about logical fallacies: S/he is old and does not publish much.' Clearly, that personal comment has no bearing on whether her or his arguments are valid. But even the more plausible 'pay no attention to the lecturer who argues about logical fallacies: S/he has no qualifications in formal logic' is a *non sequitur*: S/he might have learned formal logic from books or in the 'University of Life' or by thinking about it. The validity of an argument does not depend on the abilities or character of the person proposing it.

4. *Post hoc, ergo propter hoc* (= after this, because of this)
   Because a change has happened after some other occurrence, it cannot be deduced that the occurrence was the cause of the change. The change might have happened anyway as a result of some other unrecorded or unmentioned cause. Thus the appointment of a new dean at the Faculty of Natural Sciences is

(continued)

not reasonably attributed to the fact that a new pet shop opened in town last year. The error is implicit in the use of 'consequently' instead of 'subsequently'.
5. *Cum hoc ergo propter hoc* (= with this, therefore because of this, inferring causation from correlation)
Because measurable things seem to vary together, it cannot be deduced that any one thing causes change in the other. The increasing time a teacher spends correcting students' English shows a correlation with the number of errors students make, but it is not likely the cause of those errors. Plausibly, the errors are the cause of the time the teacher takes.
6. Confusion of necessary and sufficient conditions
Availability of water is a *necessary* condition for plants to grow, but not a *sufficient* condition. Without water a plant will not grow. It cannot be deduced that with water a plant will grow (salt water or water-logging may prevent its growth and other conditions like availability of warmth and carbon dioxide are also necessary).
7. Excluding the middle
Because a thing is not A it does not follow that it is Z. It could be any one of B-Y in the middle. Similarly, quoting a single case in which a proposition is shown to be untrue, invalid or inappropriate does not prove the universal truth of the opposite proposition. In adversarial debate and often when academics engage in personal confrontation, polar positions are adopted deliberately.
8. Circular argument or circular reasoning
This begins by asserting the conclusion as a premise and arguing that the conclusions must therefore be true. This demonstrated conclusion is then held up as supporting the original premise. Circular argument is often of the form: "A is true because B is true; B is true because A is true."

5.8   Finishing Touches, Submission and Revision

Before submission send your paper for comments to colleagues and peers that you can trust. This is particularly useful at the beginning of your research career and helps avoiding rejections and lengthy review comments. It is easy to miss something and after working on a text for an extended period you go blind for mistakes or weaknesses. Your supervisor and mentor may also be able to help, but take care that you do not get stalled, if they cannot find the time to look at your work.

List 4–5 potential target journals and sort them in descending order according to *impact factor*.[2] Discuss the list with your co-authors and potentially also with your supervisor and mentor. Start submission with the top journal,

---

[2]The impact factor or journal impact factor is an index reflecting the annual average number of citations that articles published in the last few years in a given academic journal received.

if your manuscript is declined go for the next lower. This process may take a while but it is worth optimising the impact of your research output this way. Be confident and never give up. An important reason for optimising your outlet is that increasingly mainly journal impact factors, citation indices and altimetric scores count in assessing research output and academic performance. Sadly even the contents of your papers are comparatively unimportant in such exercises, which again suggests that you should attempt to "sell" your papers to the highest ranking journals. Most of these are supported by large research communities and this is important. As long as your paper fits within the aims and scope of the journal in question, do not be afraid to aim high.

Be aware of *predatory journals* that charge publication fees without checking articles for quality and legitimacy and without providing the other editorial and publishing services that legitimate academic journals provide. Predatory journals often actively contact researchers by email and invite their contributions. Lists of *predatory journals* can be obtained from simple internet searches.

In selecting your journal, it is naturally also important to consider the audience that you want your paper to reach. Talk to peers and colleagues about the journals they read and submit to and also ask your supervisor/mentor for advice. Ideally your manuscript should be tailored to the journal you want to submit to, although this is not always possible to do in every detail. Manuscripts are often rejected on the basis that they would be more suitable for another journal. Check the journal aims and scope on the journal website and scan the editors and editorial boards. It is a good sign, if you recognise the names of the editors and editorial board members of a journal from the work you have already encountered. You can sometimes suggest handling editors in your cover letter or in the submission form, but journals are, of course, not bound to follow your suggestions and requests (British Ecological Society 2015).

Some funders mandate *Open Access* (OA) and the grant money they provide often covers the *article processing charge* (APC) required for Gold OA.[3] Some universities have established agreements with publishers whereby their staff get

---

[3] *Gold Open Access* describes a situation where the publisher immediately makes all articles and related contents available for free on the journal's website. Often the author is required to bear the costs of publication. *Green Open Access* permits the author to self-archive their work or to post it on their websites. Green OA is usually free of charge. There are also other forms of OA including *Hybrid OA*. *Black OA* describes situations where illegal access is granted on social media or dedicated sites, such as the contentious Sci-Hub site maintained by Alexandra Elbakyan.

discounts on APCs when publishing in certain journals. If you do not have grant funding, check whether your university or department has got an OA fund that you could tap into. However, if you are not mandated to publish OA by your funder, your paper will still reach your target audience, if you select the right journal for your paper. Remember, you can share your paper by email and `Twitter` (British Ecological Society 2015).

In some research communities it is also usual practice to submit any preprints of research manuscripts to an *open-access archive*, e.g. `arXiv.org`, which is hosted by Cornell University (US) and `bioRxiv.org`, which is hosted by the Cold Spring Harbor Laboratory (US). `arXiv.org`, for example, offers specialised subject archives for manuscripts in physics, mathematics, computer science, quantitative biology, quantitative finance, statistics, electrical engineering and systems science and economics. The archive also preserves all versions of an article with time stamps. In some research communities, submitting to `arXiv.org` or similar archives is the standard way of announcing to the world that a researcher has completed a paper. The original purpose of submitting to such archives was the idea of open reviewing, i.e. to give everybody an opportunity to comment on a new research manuscript. This potentially allows us to receive feedback from experts so that we can continuously improve the manuscript. Also, researchers posting preprints in open archives hope to bridge the slow review process in some academic disciplines whilst at the same time entering the manuscript into a tracking system ensuring that nobody else can "steal" their research ideas and publish them earlier. Archiving preprints openly establishes *priority*, since your work is then so widely distributed that no competitor can credibly claim to have been unaware of your manuscript. After submitting a preprint it is possible to reference the manuscript immediately and it is also accessible to online literature searches. It is often perceived that open archives invite a wider audience, since there are no access restrictions created by paywalls and many notable researchers have extensive track records of publishing in open archives. Recently the "*Peer Community in*" (PCI) was established as a non-profit scientific organisation with the objectives to create specific communities of researchers reviewing and *recommending* unpublished preprints deposited in open online archives for free, see https://peercommunityin.org. *Recommenders* in this context are subject editors appointed by a managing board and oversee the review process that may lead to the recommendation of articles that have not been published yet by or submitted to a journal. Whether you use open-access archives for publishing obviously much depends on your research community and is a good topic for conversations with your peers, your supervisor and mentor.

Scientific journals in national languages other than English will most likely disappear before long or be reduced to a very practical and minor role. Even if you are pushed for time, are at the beginning of your career or have little confidence in your work do not submit papers to such journals, as you can write an English-language paper at the same time that may in performance reviews count for five or more papers in national languages. Instead release short research notes or hold workshops for practical dissemination in your national language. A research blog is also a possibility for dissemination.

Some former national-language journals have recently adopted a new strategy and opened up to English-language articles either by publishing a mix of English-language and national-language articles or by publishing English papers only. If possible, better stick to well-established English-language journals. In this context, impact factors are also helpful for guidance. After all, if you have made all the effort to write your paper in English, definitely submit it to a well-respected international English-language journal that is likely to remain in business for a long time. Then the impact of your work is likely to be acceptable.

Prior to submission make sure that the language and every other aspect are as perfect as they can be. By writing badly or ambiguously, you antagonise readers including editors and reviewers. Any ambiguity or oversight may be used as an excuse for rejecting your manuscript or for making the revision difficult. Therefore go through several "iterations" of your initial draft and imagine you are the reviewer/editor of your own manuscript. When you do this, try to be as strict and sceptical as reviewers typically can be. Draft and redraft your work to ensure it flows well and your message is clear and focused throughout. Keep the reader in mind at all times (British Ecological Society 2015).

When sharing the manuscript with other authors it is best that all other co-authors stop editing (one-at-a-time approach), while one co-author is editing. Despite the availability of *versioning software* that track changes, simultaneous editing can lead to tremendous confusion, particularly when many co-authors are involved and the changes are substantial. Overleaf (https://www.overleaf.com/), for example, is a free collaborative cloud-based LaTeX editor for writing, editing and publishing scientific documents. Set aside enough time, as editing may take longer than the writing itself (British Ecological Society 2015). Give everybody firm deadlines, ideally both in the email subject-line and in the email body, by which they need to report back to you as the first author. The first author is responsible for coordinating all edits. S/he usually also makes final decisions. Some people find it easier to edit drafts on screen, others prefer editing printed hardcopies. Obviously all edits and suggestions should be sent electronically to the first author, but it is helpful to change the way of editing the same manuscript from iteration to iteration to minimise the chance of oversights. With different media you tend to recognise different mistakes. It also helps to change the working environment when editing. Again be sure to carefully check the *author guidelines* including the *ethical guidelines* of your target journal.

The British Ecological Society (2015) suggested going through the following list before submission:

Checklist for editing manuscripts

- Check spelling and grammar,
- Make sure all statements and assumptions are explained,
- Remove redundant words or phrases, keep your text concise and jargon-free to avoid diluting your message,
- Check that all abbreviations have been explained on the first use,
- Make sure all funders are clearly mentioned in the Acknowledgements and that all people who contributed in any way are acknowledged,
- Key words should be consistent, evenly spaced throughout the text and placed at key points in your manuscripts, e.g. subheadings.
- Finally, make sure you have specifically dealt with the hypotheses set out in the Introduction.

The online submission can easily take several hours. Plan for sufficient time and make sure nobody can disturb you during this time. The *online submission system* (e.g. ScholarOne Manuscripts, Editorial Manager) of every journal is different and you may discover that you still need to prepare a few smaller documents (e.g. highlights, cover letter, author credit statement, statement of conflicting interests etc.) whilst submitting your manuscript. Take a deep breath and do not get upset then. In the worst case you can save your submission progress and continue later or on the next day. Although these systems are fairly easy to navigate, be prepared to enter quite a lot of information during the submission process. The system will prompt you for all necessary information and you can always contact the journal's editorial office, if you have any questions during the process. Only submit your manuscript for consideration to one journal at a time, otherwise you will be breaching publishing ethics (British Ecological Society 2015). The online submission system usually also asks you to provide the names and details of potential reviewers. Obviously these should not include colleagues from your own institution and it is good practice to put forward researchers from different parts of the world. Sometimes you also need to explain why you suggested particular reviewers. It saves time to consider suitable reviewers in advance so that you can quickly move on in the submission process.

Some journals require you to compile a short list of *research highlights*. This list is fundamentally different from the summary: Here you condense the major scientific selling points of your manuscript. They need to be very short and written in the style of an advert for commercial products. Be creative and take your time. This is not an unimportant step of manuscript preparation. In the same way it is important to write good *cover letters*. For highlights there are usually good examples in the papers previously published by the journal you selected as outlet. Take a look at a few of these examples and then write your own highlights modelled on the examples you read. Examples of cover letters are harder to find in the internet. A great cover letter can set the stage towards convincing editors to send your paper for review. Write a concise and engaging letter addressed to the editor-in-chief, who may not be an expert in your field (British Ecological Society 2015). You should refer to the title of your paper, mention the authors, the journal and state your key reason why your manuscript is important and should be published in that journal. Sometimes it is required to state that your paper is not under review in another journal and has not been published before. Here is an example from one of my earlier submissions (obviously cover letters should be printed on an official letterhead and include all relevant addresses, dates etc.):

Dear Editor,

We are submitting the manuscript entitled "Unravelling the mechanisms of spatial correlation between species and size diversity in forest ecosystems" by Pommerening, Zhang and Zhang for possible publication in *Ecological Indicators*.

In our research, we analysed the causes of spatial correlations between species and size and thus our findings contribute to understanding the natural mechanisms of maintaining plant diversity. This understanding is essential for mitigating the loss of plant diversity with ongoing climate change.

We are confident that the methodology, the results and the new insights of this study are of great interest to the international biodiversity and ecological community. This manuscript is an original manuscript and not under consideration elsewhere.

I look forward to hearing from you.

(continued)

Sincerely,

Arne Pommerening
(Professor)

The journal editor, often a senior academic, will briefly scan your manuscript and decide whether it might be worth publishing. The majority of papers is returned to the authors without review at this point. This is part of an initial step of quality control and returning manuscripts that do not match the theme of the journal or have little potential is referred to as *desk rejection* (Ritchie 2020). Peer reviewers act as a bridge between the journal and an author and one of their functions is to ensure that all aspects of the manuscript – the research it presents, the coherence of the document and its authority and persuasiveness – are all up to the standard and quality required by the journal's editorial team. Often it is true that if the reviewers cannot understand something the authors have written, it is likely the journal audience will not either (Hopkins et al. 2020).

Therefore when the review comments come in, never get upset about them. Remember, none of the comments are personal and reviewers often do this work in their spare time and may just have been tired when writing them. After all they are just human beings like us and also happen to be prejudiced from time to time as are we. Many are pushed for time and do not read the texts they review properly. Then they can occasionally arrive at wrong conclusions. This is not ideal but also not a big deal, you can gently correct their views in the response document of your revision. Learn from this experience and when you carry out reviews yourself, try to make a better job and to be more polite and encouraging.

Journals usually ask for a letter outlining your *response to the reviewers*, i.e. describing how you have dealt with the review comments. This is also sometimes referred to as *rebuttal*. Writing this document is an art and requires special care. Novices (particular PhD students, postdocs) should enlist the help of their supervisors and mentors when writing their first responses to reviewers. This is an important part of the research education experience. For some researchers it works best to start the revision with drafting this document. There is no standard or even any particular rules for writing such documents, however, you should reply to all comments politely. Not answering comments or being rude in your response will decrease the chance of acceptance and harm your reputation (British Ecological Society 2015). Often the structure is more or less a matter of tradition in the respective research communities. A template of an effective structure of a response to reviewers that I have used a lot is given below:

Response to the Review Comments

Manuscript number:

Authors:

Title:

First, we wish to express our gratitude to the editor and to the reviewers. Their comments were a great help in improving the manuscript. All review comments and our responses are listed in the table below.

Reviewer #	Comment/line in old manuscript	Response	Line in revised manuscript
1	45–67	Thank you, we followed your advice.	76–88
1	120–135	This diversity aspect has been much undervalued in the past as noted by eminent plant ecologists. We have re-emphasized this point in the revised manuscript.	123–145
1	…	…	…

Only argue with reviewers if absolutely necessary, e.g. where you feel the reviewers have not made an effort to understand parts or all of your work. If you disagree with certain comments, disagree politely and with evidence. If things cannot be resolved, explain that to the editor, as reviewers sometimes may try to push their own agenda (British Ecological Society 2015). Otherwise take their advice seriously and welcome revisions as an opportunity to improve your paper. However imperfect the peer-review system may appear, the simple truth is that we all need this kind of feedback to enhance the quality of our work. All papers usually benefit from and improve in the review process. When writing your response document try to adopt a positive writing style so that even where you disagree your response sounds as if you agree.

If a review was (partly) negative and gave you grief, give it a rest for 1–2 days, let it settle. Ignore any bad language or impoliteness (consider that as "noise" just as Shostakovich did, see Sect. 9.2). Rather focus on the positive things, on pieces of good, constructive advice. Negatively worded comments may have a constructive core that you can work with for the benefit of your paper. Even if the editor finally rejected your paper, use the advice given to modify your manuscript before submitting it to another journal on your list. However, do not follow all suggestions, particularly if after giving them a lot of thought they seem odd and out of place. The opinion of one reviewer may not be shared by another and it can occasionally be that one reviewer unnecessarily sent you on a wild-goose chase.

Occasionally the peer-review process can, of course, be lengthy and even painful. Then we need to take a deep breath and muster all our strength. Not seldom our papers or parts of them are misunderstood, although we made a great effort. Some reviewers act on a hunch and subject editors follow them without verifying the reviewers' opinions. Here is an interesting account that may encourage all of us to endure.

> **An experienced, senior colleague once wrote about his recent publication:**
>
> "I am just reading a page proof for a publication. Thought the little story may encourage:
>
> The paper is about the contribution of mycorrhizal hyphae to soil organic matter. It is new and controversial. The oddyssey......
>
> Sent to Nature – rejected, not of general reader interest.
>
> Sent to Science, out to review, a big success. Reviewed by 3 refs, two think it is great, one slams it. Rejected. We work on the data. Think the nasty reviewer had made a big error. Resubmitted to Science. Two think it is great, one slams it. Rejected.
>
> Submitted to New Phytologist. Two think it is great, one slams it. Rejected, but please resubmit. Resubmitted to New Phytologist. With extensive answers to referees' comments. Two think it is great, one slams it. Rejected.
>
> Sent to Plant and Soil. One thinks it is great, one slams it. But this time the comments are the most bizarre ever seen. Rejected, but please resubmit. Resubmitted to Plant and Soil. With extensive answers to referees' comments. ACCEPTED!!!!!!
>
> A total process of over 2 years, 6 versions and 17 reviews."

Papers get rejected all the time. If your paper gets rejected, it is absolutely not a personal failure. Keep in mind that feedback including that associated with a rejection is just another person's opinion on what you have done, not on who you are and it is up to you to decide what to do with it. If you are unhappy with a reject decision, 99.9% of the time it is best to move on and try another journal (British Ecological Society 2015). If you are sure that your manuscript was only superficially read and because of that misunderstood, submit your paper to another journal without modifying too much. And do not worry, your manuscript will be accepted and published by one of the journals on your list eventually.

Once your manuscript has been accepted, the hassle is unfortunately not over. Most journals have a specific post-acceptance workflow that comes with a specific set of rules. Whatever they are, in a short while you will receive the proofs of your accepted manuscript. Then you really need to take the time to read and review these properly. They are usually full of mistakes. Typesetters are unfortunately under huge pressure these days and they mainly rely on the authors to correct their work. Some mistakes are, of course, of your own making and it is natural that you tend to see these oversights only now. After typesetting they appear in a different format and style and apparently this helps to spot previously undetected problems. Equations and tables are particularly prone to typesetting errors. Distribute the proofs to your co-authors and ask them to read the proofs carefully and independently. The more people separately look at the same proofs the better the final results. Give your co-authors a deadline and collect all problems they have found. Then send a coordinated response to the journal. If there were many mistakes, ask the journal kindly for another set of proofs to review. This work is really important, since it is a shame for both the authors and the journal in question when publishing an article that is full of mistakes after all the trouble both parties went through.

5.9 Promotion and Celebration

We are well advised even to do more than publishing by actively "marketing" our papers on Twitter, Instagram, Facebook, on websites and perhaps by emailing them along with a short summary to potentially interested colleagues. This may sound strange, since the traditional view was that a research paper should make its own way to convince others by its inherent quality, see end of Section 5.4. However, there are so many publications these days that it is hard to keep up to date. Many colleagues are grateful for an alert and this is also a great opportunity to invite feedback and to have a little discussion on your latest paper. Who knows, perhaps this will inspire new research questions or trigger new cooperation. Send your published paper to interested colleagues and include a reference to it in your email signature. A short summary or a few highlights of your research included in the email text may inspire interesting follow-up discussions with the recipients. Also use your personal websites to promote your publication activities. Advertising your paper in these ways may support your citation index and interested colleagues are alerted to the existence of your work.

Blogs and podcasts are also effective ways of promoting and disseminating research. You can, for example, write a blog telling an intriguing background story "behind the scenes" of a particular article you recently published, e.g. how the idea was devised and what happened during the fieldwork or the analysis. Here you can also share any amusing anecdotes. Visitors of your websites will enjoy reading these stories.

French researchers have recently founded a new *outreach journal* with the objective to publish outreach versions of articles published in scientific journals. These outreach versions should be written in a highly accessible way and are reviewed by school children, see http://journal-decoder.fr.

Universities and similar research institutions typically have press or media offices who can support you in telling your story. This is an important part of their work and they have expertise in dealing with the media (British Ecological Society 2019). Make sure you visualise your research story with interesting photos (from your fieldwork perhaps) or graphical abstracts. Avoid jargon, acronyms and measurements that the public may not understand. If you are referring to scientific terms or statistics, make sure to explain and put them into a relatable context.

The British Ecological Society (2019) suggested the following general guidelines for dissemination and promotion of research:

> **Be direct:**   'I investigated…' is better than 'an investigation was conducted…'.
> **Be active:**   'We measured each bee' is better than 'each bee was measured'.
> **Keep it jargon-free:**   Do not use technical terms when plain language works just as well.
> **Be clear:**   'It's important citizen scientists know how and why they're counting bees this way' is much better than 'the importance of understanding this methodology by non-scientists acting as data-gatherers can by no means be underestimated'.

(continued)

Respect your audience: Do not assume they know what you know...but do not think that means they cannot understand. This is true when talking to colleagues and it is true when talking to everyone else.

Be involved: Enthusiasm is contagious! 'Your story' is more interesting than 'a story'.

Make it relevant: Why is it important? How does it affect that member of the audience personally?

Make it relatable: Give a frame of reference the audience can immediately comprehend, e.g. 'the size of your thumbnail'.

Finally and most importantly, take time to celebrate every published paper. Do not take your success for granted or simply move on. However small the paper may seem to you, it is a certain measure of success and you have reached an important milestone. Invite your partner to a restaurant or open a bottle of wine (or whatever you fancy) at home. You truly have deserved this treat and it is important that you have it. At some institutes, it is a nice tradition for the first author to bring cake for the whole team on such occasions (Gosling and Noordam 2011).

5.10 Proposal Writing

Given the need of most universities for winning external funds, it is crucially important for every academic to engage in grant proposal writing. This is possibly the second most important activity after writing scientific articles. Funding records are often requested in job applications and external grants offer a unique opportunity for building up your own research group and thus to live your own research vision. Since universities and similar research institutions are notoriously short of funds, external funding is crucial and funding records also play a major role in promotions. Hopkins et al. (2020) summarised the benefits of successful grant applications:

- Greater grant income to the institution,
- Increased research outputs (journal articles, conference participation, impact etc.),
- Increased potential for knowledge exchange and commercialisation of research,
- Enhanced performance in national academic performance reviews,

(continued)

- Heightened institutional prestige and esteem,
- Ability to attract and retain the best researchers.

Usually every university campus has a *grant office* with dedicated staff that is employed to relay information on new calls and advise on grant applications. Make sure you subscribe to the newsletters or emails they issue and stay in contact with them. You can and should, of course, also actively search for national funding opportunities yourself. Also ask around in your networks, whether they know of any international funding opportunities that you may team up for together.

Proposal writing is also a variant of scientific storytelling. Proposals are in fact just one type of the many different texts we have to write in our careers. Usually those who successfully can write research papers, are also able to write successful proposals and vice versa. The principles are much the same and even the structure of proposals often does not differ too much from that of research articles. Notable exceptions have recently been introduced by web-based proposal software that only accept plain text so that including equations and even figures is made difficult if not impossible. Similar to research papers successful proposals tell an intriguing story and are about solving an important question linked to some theory. Both activities require researchers to set for themselves clever tasks that are likely to lead to intriguing results.

When planning your time remember that successful proposals often require nearly as much time as papers. Therefore carefully balance these two activities and cooperate with others to share the workload. Successful proposals require a solid list of publications as pre-requisite. A good track record of successful proposals is important to your career, but published papers are more important. Whilst with the necessary skills, experience and perseverance you can potentially get nearly every paper published, by far not all of your proposals will be accepted and they rarely get a second chance either.

Since there is an obvious similarity between paper and proposal, try to re-use some materials of failed proposals to write a research paper afterwards. There is sometimes also a possibility to distribute failed proposals that you submitted in your country within your international networks. Your colleagues may be able to submit them after some moderate modifications in their home countries and can include you as a foreign research partner.

Also, much of what we discussed in the context of writing research papers applies to proposal texts as well. For proposals it also pays off to team up with experienced, senior colleagues and to seek the advice from peers, supervisors and mentors.

Writing a proposal and organising a new project is a major operation. It is therefore a good idea to secure administrative help from your department or faculty, e.g. from a dedicated research or grant office. Definitely get someone to check the finances and budgets, as this is usually beyond the training and education of researchers. In most cases university grant offices make semi-automated MS EXCEL spreadsheets available to ease the burden of devising research budgets. However, double check with an economic officer to ensure that all your calculations are sound.

When writing your own proposal, it is good to know what the most common reasons (or pretexts) for rejecting research applications are (Hopkins et al. 2020):

- Proposal does not fit with funder's strategic priorities or remit,
- Application was rated as excellent but owing to limited funds, even the best-ranked bids were not all supported,
- What you plan to do is not sufficiently topical or exciting or cutting-edge or interesting or high priority,
- Methods are inappropriate for the topic or mundane or out of date or beyond your capability to carry out,
- Budget is unrealistic or does not represent good value for money,
- The bid was in some way incomplete, biased, unsubstantiated, poorly argued or badly written.

With these arguments in mind, you can try to write your proposal text in such a way that they do not apply. However, the process of reviewing research proposals usually is far less transparent than that of manuscripts submitted to scientific journals and fair feedback is hard to obtain.

Increasingly the question whether you will be awarded funding depends on whether the general research topic is popular with national research councils but also on *gender balance* and other factors. There is much more politics involved in grant applications and grant distribution than in publishing. To avoid rejection, depending on country and your research field, it is sometimes necessary to get collaborators from fields that are quite different from yours involved in your proposal or even to write about a topic that is not fully within your main field of expertise. Particularly for large and strategically important proposals it can be a good idea to arrange dedicated meetings or even retreats to make the process of writing easier.

In some situations it helps to secure 3–5 letters of support for your proposal from important stakeholders. They can be appended to the main proposal text and may help to convince the reviewers. Funding bodies often require applicants to include plans for data management (see Chap. 10) and for research dissemination. Field trips, where your field plots can feature prominently, workshops and conferences are good ideas to include in dissemination plans.

When designing your project try to organise your proposal milestones and deliverables by publications. Avoid promising lengthy reports to be written. If necessary at all, keep project reports short and append your papers/manuscripts as accomplishments.

Sometimes the industry that benefits from a certain research field offers scholarships and funding of research projects. This can be a good source of income for your research group. However, you need to check whether the funders give you sufficient freedom in terms of research objectives and methods to pursue academically rewarding research questions. Also the applied objectives of the project must provide good material for publications. In some countries, universities appreciate grants from research councils more than grants provided by the industry irrespective of the amount of money.

As part of personal record keeping maintain a list that includes both successful and unsuccessful grant proposals to include in your job applications. It is important that all your efforts are listed in that record.

6

Scientific Presentations

Abstract Despite the advances of modern communication technology, researchers regularly need to meet at conferences and workshops. Web-based seminars and even video conferences have now increasingly become an alternative and here similar guidelines apply as for the "real" meetings. Research meetings are an essential part of academic life. Meeting someone face to face does makes a difference and is good for networking.

In the same way it is important to give research talks at international meetings on a regular basis. When you sign up for a conference always offer a talk or a poster presentation so that people notice you and your work. This is good training and also helps to get in contact with people and you receive feedback on your research. Actively taking part in discussions also makes people remember you and come back to you. Introduce yourself to others during the social events and talk to as many people as possible.

6.1 Preparing Your Presentation

In your scientific talk, plan to present the most important results of your research. Do not give too many technical details, as they may distract your audience and bore them. Focus on the highlights of your work and the big picture of what you found in your research. Ideally as a basis of your talk you select one of your papers that has recently been published. Then everything has been thought through and you have already produced the figures. The review comments received in the peer-review process may help you to anticipate

A. Pommerening, *Staying on Top in Academia*,
https://doi.org/10.1007/978-3-030-65467-2_6

questions in the audience. Also there is then no chance that others in the audience might quickly repeat your experiments and publish their results before you do. Hopkins et al. (2020) offered a self-reflection designed to gauge the needs and expectations of your audience:

- Who will you be presenting to?
- What is the purpose of your presentation?
- What does your audience hope to achieve from your presentation?
- Are the members of your audience specialists in your field?
- If they are not specialists in your field, what will they need to know about your research?
- What technical terms can you replace with commonly used terms or phrases?
- If you must use technical terms, how will you explain them?
- In view of your audience, what can you say about your project that will excite them?

Dedicate one slide to every single thought. Each slide takes approximately 2–3 minutes to talk to. Based on this you can calculate the number of slides you can present during the time that is allocated to you. As a rule of thumb for an upper limit, try not to use more than 15–20 slides in your talk (scheduled for 20–45 minutes presentation time). Definitely do not get tempted to compensate for too many slides by speaking more quickly. On the contrary, your pace should be really slow and constant during the whole presentation.

Slides and posters should include as little text as possible. Processing text on slides that change comparatively fast is very hard to follow and tiresome for anyone in the audience. Slides are visual media, i.e. pictures, graphs etc. play a key role. Give most of your information orally so that the audience can focus on what you are saying. In ideal terms the slides only help you to remember and illustrate what you intend to say, but you should not use them to read something out aloud. Therefore keep your slides simple with only few elements on them and graphs as large as possible with really big axes labels. Then talk freely about what is not on your slides. Obviously this requires good preparation.

When preparing make sure you meet the audience in "their world" at the beginning of your talk, i.e. you start with something they are very familiar with, perhaps something from everyday life, and then gradually introduce them to your thoughts and concepts. Be mindful of their ignorance and level of knowledge.

It is good practice to keep an appendix with "reserve" slides for details that are not included in the main talk. Use those in the discussion to answer method questions should someone raise them. It looks very professional, when people realise that you have anticipated some of the questions. You can also proactively place someone in the audience who will then ask a specific question so that you can show a slide from your reserve that you believe is very important.

The layout of computer-based presentations can be very different on different computers depending on software versions and operating systems. Try to be independent of such surprise effects by using LaTeX beamer presentations or similar software producing animated pdf files as output. Make sure the conference computer has Adobe Acrobat installed for viewing your presentation pdf.

Flashcards are good for learning and training your research talk prior to delivery. However, never use them in the real talk, as they may distract you, when you are nervous.

Many research talks are being recorded these days and you can watch them on YouTube. This is also a good way of picking up interesting ideas for your scientific talk and to improve your presentation skills. Inaugurational lectures are particularly interesting.

If you are a junior researcher, you may want to start with a poster presentation rather than a talk. PhD students typically present posters at conferences. This is not less demanding than the preparation of a presentation and careful design plays an even greater role. During poster sessions there is a great deal of personal interaction and they can be extremely rewarding. Naturally you cannot stand by your poster at all times, so be sure your poster is self-explanatory. Print a stack of A4-sized sheets of your poster and place them near the poster so that people can take a copy with them. Also it is a good idea to put your photograph somewhere on your poster. This helps people to find you later on, if they want to discuss your research with you (Gosling and Noordam 2011).

6.2   Showtime

For training the delivery of conference talks use lecture rooms that are similar to the conference venue. Ask colleagues to sit in and to listen, particularly if you have little experience or confidence in giving conference talks. You can, for example, use a slot in your department's seminar series to give a mock talk for practising. Take your colleagues' feedback seriously and, again, try not to speak from written notes.

On arrival at the conference make sure you can check out the conference venue prior to delivering your presentation. Ensure that the required software is available on the presentation computer and your file works well. Alternatively, if you are allowed to use your own computer, make sure you brought all the necessary cables and adaptors. Also take time to test the remote control available in the conference room, as it is easy to confuse the buttons, e.g. to press 'forward' instead of activating the pointer. It can also be a good solution to bring your own remote control along that you are familiar with.

When delivering conference presentations *never* speak longer than the time allocated to you. Using 'overtime' once your official time slot has ended is considered a serious offence, since you are robbing other speakers of their time. Therefore take enough time to train yourself in delivering the talk and particularly in finishing on time without using a stop watch during the actual delivery.

Try not to fall into the trap of "lecturing" in your scientific talks. Remember you usually give your presentation to experienced specialists who may know more than you. Lecturing may annoy them and prevent good discussions.

From time to time re-address people directly to secure their attention (e.g. "Dear colleagues, we are now moving on to …").

Use each talk for multiple events

Preparing a scientific talk takes a lot of time and there are many different occasions where we are asked to give talks. If these occasions include quite different audiences and locations, why not delivering one and the same talk several times?

Drop a few jokes in between and keep smiling. The British Ecological Society (2019) suggested that it helps to smile before speaking, as smiling lifts the tone of your voice and make you sound more engaging. Let the audience feel the passion you have for your research. If you are nervous, try not to show it. There is nothing to be afraid of and everybody started somewhere. Look at the audience (i.e. move your eyes to the left, centre, right, front, middle back of the venue) while delivering your talk, address them and avoid facing your slides on the screen instead. Never turn your back to the audience. This is in fact a common mistake. If you need to point out something on the screen, do that, but before you continue to speak turn to face the audience again. Also avoid looking at the same person all the time.

Moderately use gestures to support your statements and for securing attention. It can sometimes also be very effective to use props to draw analogies or emphasize your points, e.g. to flash your mobile phone when talking about signal transmission. Speak slowly and clearly and give a relaxed impression. If something goes wrong, turn it into a joke, but do not get upset, particularly not when "on stage".

If possible, do not allow questions to be asked during your presentation and kindly ask the audience to wait until the end of your talk. Never take questions or criticism personally or consider them as a threat. Thank everybody lavishly for their question and remark, even if they included a criticism. Stay calm and cool. Clearly nobody expects you to know everything. Give every comment credit and state that you wish to take suggestions into consideration in your ongoing research. If a person asks more than one question at a time, take notes so that you can remember them.

If you have not fully understood a question or need time to consider your answer, re-phrase the question and ask the person who raised the question whether you have understood correctly.

Needless to say that it is a good idea to attend conference dinners, field trips and receptions to meet as many people as you can and to have informal discussions. Breakfast at the hotel, coffee and meal breaks are also ideal for this activity. During these occasions you hear about many useful things such as trends, job opportunities and research strategies that people attempted but failed (Gosling and Noordam 2011).

Once the conference has passed and you are home again, why not follow up some of the formal and informal conversations you had at the conference by emails? Perhaps you can exchange publications, arrange a mutual research proposal or write a paper together with the people you met at this research meeting. If it has not been published already, most importantly make sure that your conference presentation will end up in a good scientific journal. Sometimes the organisers of a conference have arranged a special issue with a particular scientific journal. You can aim at submitting to such a special issue, if the outlet is good, but avoid conference proceedings that are considered as "grey literature" and do not really count in performance reviews.

6.3   Chairing a Conference Session

Another way of getting known is to chair conference sessions. Chairing a session is also an important skill to practise. You need to introduce the speakers and also to ask questions, particularly, if the audience have none.

If you are charged with organising a whole session, first identify the leading authorities relating to the topic of the session. Try to secure their participation early on. If you are not so experienced, discuss this issue with your colleagues and with leading individuals in this field that you happen to know. Then fill the gaps with local colleagues and with people who have registered for the conference.

Ask for short cv's of the speakers or better find out about their careers in the internet so that you can properly introduce them to the audience. Manage the sessions in such a way that no speaker takes up more time than is allocated to his or her slot. Be strict about this and use signs as warnings to show that only 5 and 2 minutes of speaker time are left. If you fail to stick to these rules, the conference schedule may seriously get out of control. It is crucial to leave enough time for discussions, otherwise many talks may pass by unnoticed.

Perhaps you find the time to meet with the speakers of your session for lunch or dinner? Seize this opportunity, since this is beneficial for networking and you may develop new ideas during the informal conversations at the table.

7

Research-Driven Teaching

Abstract Teaching is an important aspect of academic posts. Making sure that the next generation of researchers and practitioners is properly educated is both a rewarding and an inspiring task. Including one's research in teaching and allowing new research questions to be inspired by student questions is even better. However, since academic tasks are on the increase and new challenges need to be faced and accommodated in busy schedules, it is also important for teaching to be as effective as possible so that research activities are not compromised.

Teaching needs to be considered in the context of research and likewise teaching is often strongly connected with our research activities and results. This chapter offers a few points that may help you to find a good balance between teaching and other academic responsibilities.

7.1 Teaching Organisation

Biggs and Tang (2011) pointed out that teaching is nowadays much more student-centred than it used to be and in their book they described what they referred to as *outcomes-based teaching and learning* (OBTL). This shifts the focus from the lecturer to the learner and specifically involves defining learning outcomes that students are meant to achieve when academics address the topics they are meant to teach. At the same time it is also important to understand that high quality teaching at universities should also be research-oriented and considered in the context of research.

© The Author(s), under exclusive license to Springer Nature Switzerland AG 2021
A. Pommerening, *Staying on Top in Academia*,
https://doi.org/10.1007/978-3-030-65467-2_7

The *Bologna Process* was originally intended to facilitate credit transfers between institutions in different countries and the Process has also much influenced our research education. As such teaching has become much more formalised and rigid compared to previous decades. The Bologna Process has increasingly encouraged the reflection of what is happening worldwide (Biggs and Tang 2011). This trend emphasises that we are all part of a changing world in terms of university teaching and ongoing pandemics may now accelerate this process.

You should always take an interest in good quality teaching, as this is a crucial part of academic life and culture. Teaching also fuels your imagination and student feedback has often led to excellent research papers. Whenever you feel enthusiastic about a new teaching idea, write the idea down and come back to it when your research schedule allows you to do so. After all in research-oriented teaching, our research results feature prominently in our courses.

In some countries, universities try to free up time for established scholars by delegating their teaching to teaching assistants. Though it can be tempting to accept such an arrangement, I believe that teaching is crucial to good research outputs and increases job satisfaction. Teaching is also a way to ensure that our ideas make their way in the world and that our research vision is continued. Also, students surely want to meet the famous professor they have come for to your university in the first place. Then the teaching experience is much more valuable for them, if the professor includes his own research results in the lectures so that the students are invited to share the scholar's research work.

It is wise to organise teaching and teaching preparation so that there are no times during the year when you cannot work on research papers and research proposals. This implies that you need to start preparing well in advance, perhaps a year before the start of a new teaching module by setting aside 1–2 days per week for the preparation. This, of course, also requires good communication with those who assign teaching loads at your university/department and particularly with your line manager.

For many researchers it does not work well switching different tasks during the course of a day. Particularly when concerned with research, even small disturbances can severely interrupt creative thought processes. On the other hand it is often not a problem to be interrupted when carrying out administrative tasks such as filling in forms. You need to test yourself, but blocking your time by setting aside dedicated days when you only concentrate on research, teaching or administration, may work better for you. The change in tasks that comes with different days of the week also often increases motivation, as you can take a break from one activity for one or two days.

> **Tip**
>
> However you are inclined to organise and plan your time, it makes sense to ring-fence at least *two research days* per week when no one is allowed to disturb you, even not your family. This will secure a constant flow of research outputs. Switch off emails and the telephone. Be strict about this, also to yourself. Obviously it is important to secure the approval of your superiors for this, but with increasing willingness to allow people to work from home this should not be a problem.

To help share your teaching load it can be a good strategy to invite colleagues from other universities and from abroad. Some colleagues may have a much smaller teaching load than you and are glad to help out and to share their insights at other places. Retired academics with great teaching and research experience are often willing to step in. At the same time you can have research discussions with these people and the students love to see other faces and to listen to other voices from time to time. International and regional grants often support such teaching exchange.

When preparing teaching material, if possible, do not engage in writing substantial lecture notes (sometimes referred to as 'scripts'). This activity requires valuable time and dedication which you should spend on research. In performance reviews, it hardly happens that excuses are made for bad publication records and that you are praised for preparing additional teaching material. If in doubt rather write an (English language) textbook instead of lecture notes. This also has the added benefit that students can make valuable experience with scientific academic textbooks, something that is essential in their career.

> You can make a great difference to students' teaching experience by making all your slides and additional material available on websites dedicated to the modules you teach. Often universities have purchased a commercial teaching platform (e.g. Blackboard, Brightspace, Canvas, Moodle), where each module has dedicated space for uploading teaching material. If this is not the case or access is difficult, why not use your own private website or your GitHub account (https:// github.com/, see Chap. 10) instead? Also recommend textbooks in your teaching materials, when you introduce your module. For obvious environmental reasons try not to spend excessive time on printing/photocopying material. Encourage

(continued)

students to accept digital copies of your learning material and to take those from the specified module websites.

If you teach English and national-language classes on the same topics, try to prepare all your teaching material in English to save time and effort. In the national-language classes, you can then orally teach through the medium of the national language using English-language slides. This is also a good experience for students in classes taught through the medium of the national language, as they learn the technical terms both in their own language and in English at the same time. Obviously this needs to be discussed with the module organiser and/or teaching director, however, given current national and international developments the chances are high that your concept will be accepted, since it is efficient and has a number of other advantages.

Always be open to teaching reviews and teaching observations, when colleagues or pedagogic professionals offer or want to sit in your lectures. This is now common standard and good academic practice. Such observations usually lead to very positive discussions and to an improvement of your teaching. In fact, this practice can be considered an important part of academic mentoring, see Chap. 1. Often the observer also takes messages home from your teaching (see reverse mentoring in Sect. 1.7). Perhaps you even proactively invite colleagues to sit in and to give you feedback, if there is no such system at your university.

It is a standard and laudable practice by now that students assess university teaching using formalised surveys, often in the format of questionnaires, issued by the university. These surveys are then analysed by dedicated staff and the results sometimes play a role in the teaching accreditation of universities, but more importantly they provide valuable feedback. Such feedback can be very useful for improving our teaching. However, these student assessments usually vary a lot from year to year, mostly depending on the varied structure of each class we teach. They can be particularly negative when your module runs for the first time. In the same way as you should not be upset about review comments on your submitted manuscripts (see Sect. 5.10), you should not worry too much about uncomfortable student comments either, particularly, when you really made an effort. Again, be grateful for any comment, since such feedback indicates that the students were interested in your teaching, and focus on the positive side of the comments you receive, even if they sound negative. To obtain an unbiased opinion about your teaching it is best to consider the sort of average student-review results from 3 to 5 years, since extreme assessments usually cancel each other out over the years.

If you teach multi-national or dyslexic students, you may want to offer that your lectures are recorded. The digital records of your lectures can then be posted on the module website and the students can hear your explanations again whilst revising the slides. At the same time recording your lectures can also be used for distance learning efforts, particularly during times of pandemics. Currently this is more important than ever. Do not be shy, many people find it embarrassing at first to hear their own voice and to face their imperfection when lecturing, but this is for a good cause. Recording lectures is also useful for part-time students and students who missed your lectures for other reasons.

At the same time many lectures given by others are being recorded these days and you can watch them on YouTube and/or ask your students to watch them. Using such material is also a good way to pick up interesting ideas and to improve your own teaching skills by watching how others do it.

Some academic societies such as the American Society of Plant Biologists (ASPB) have engaged in providing *teaching repositories*. In doing so they address the stress and pressure busy educators face when devising course material. Interested individuals should provide a short essay introducing each topic, PowerPoint slides and suggested readings, see http://www.plantcell.org/content/teaching-tools-plant-biology. This is a good possibility of exchanging teaching tools, particular during the challenging times of pandemics and associated online teaching, and may give academics with high teaching loads some well-deserved recognition.

Be careful with volunteering too many modules or courses: The offer is usually taken and it is hard to lose them again in subsequent years when you realise that your teaching load is too high or you need to free up time for other academic activities.

As a result of international trends we increasingly offer modules to non-specialists. This is particularly common at postgraduate level (MSc and PhD). This trend requires special care in the preparation and delivery of modules, as we cannot assume all participants to have been exposed to the necessary basics. Our modules therefore need to be sufficiently self-contained whilst still offering enough new material for students who have already covered the basics. Openly discuss this problem with the students and involve them in finding the best solution. One possibility is, for example, that more advanced students help newcomers to the subject area by running small tutorials. Clear benefits of such mixed classes typically include that students with a different background often ask interesting, unorthodox questions that encourage the whole class to think 'outside the box'. Also, students with a different background often contribute their expertise from the field they studied before.

7.2   Teaching Style

University regulations permitting give your students the choice to attend or not to attend your lectures, as this is true academic spirit. Tell them that you are not upset about their absence, however, that they are then fully responsible for their learning success or failure. It is part of higher education at universities that students learn to make choices and best use of their time. To enable this they must be allowed to make a few mistakes so that they can learn from them. The skill to make good decisions will come handy later in life and is also required in future jobs. However, recommend (but not demand) the participation in field trips, field, lab and computer practicals, as students cannot study their contents from books.

Do not lecture in lengthy monologues, rather adopt a seminar style wherever possible and involve the students by asking questions and by encouraging discussions. It is also possible for parts of a course or module to ask students to orally report to the class about a research paper or a specialised topic. This potentially introduces a valuable element where the students need to become active themselves. Such activities help to overcome sleepiness and make students apply what they have learned. Applications of knowledge usually are better remembered than simple learning by listening or studying texts. Praise and encourage every student contribution.

It can also be useful to adopt elements of the *flipped classroom* teaching strategy (Nouri 2016; Tomas et al. 2019). This technique is focussed on student engagement and active learning by moving activities that traditionally were considered assignment work into the classroom. As part of this, students have to listen to recorded lectures and consult textbooks *before* they come to the class and the focus of class work is on discussions and group work. This technique is often perceived to result in better learning outcomes for the more active involvement of students and because topics can be explored in greater depths in class. Also the lecturer is perceived to be in a better position to deal with mixed levels and student difficulties. However flipped-classroom teaching can go wrong when the students are not prepared and the adoption of this teaching

(continued)

method should not be used as an excuse for a lack of preparation on the lecturer's part. It is probably best to use a balanced mix of traditional, seminar and flipped-classroom techniques. A good idea can, for example, be to put a small *literature seminar* at the end of a course where small groups of students have to present a research article to the class relevant to the topics discussed in the course, see Sect. 5.3. This encourages active learning and gives the lecturer some "breathing space", particularly when the course is new and there was not sufficient time for preparing the whole module in one go.

Speak slowly and in a clear and firm voice. Avoid any artificial style of talking. After a few years time some lecturers develop the habit of speaking in a particular way in class that is fundamentally different from their normal way of speaking. Watch yourself and try to avoid this, act normal. When lecturing keep watching the eyes of your students. When you see they get tired, change your lecturing style and try to be more enthusiastic and emphatic. Regular breaks, small activities and the provision of coffee/tea may also help. Walking up and down the class room is also helpful to regain attention. Never raise your voice, rather reduce volume and rely on self-discipline among students.

Computer-based presentations can sometimes be tiring, especially if used excessively for hours. It is also possible to cheat with such presentations, for example, google for *PowerPoint karaoke* to get a feeling for the potential misuse of presentation software. This is also important to bear in mind when setting assignments that involve such presentations. LATEX beamer presentations are often a good choice, since they follow certain design rules, although similar problems exist. Also consider using traditional white/blackboards. *"Chalk talks"*, where you write as you go along in your lecture, are often better received by students because of the slow pace of delivery of information and have a somewhat less tiring effect.

It is interesting to note that we generally do not fascinate our students with the most elaborate slides but mainly by your own enthusiasm. When students see that your heart is in the lectures and that you are passionate about the subject, they will love the topic and feel encouraged to learn.

Updating your teaching materials is necessary and usually you sense yourself when the need arises to do so. However, updating can take a lot of your time, so do not do this every year and do not update everything in one go.

As mentioned before, including your own research results in your lectures makes them really lively and is truly inspiring for students. They need to feel that they are at university and not at school. This way you also receive feedback from the students on your research and in the process of discussing your research often new ideas and research questions pop up. Therefore take notes during the lectures, which slides worked well and which did not. Write down interesting student comments and questions. Remember to come back to these notes after the lecture to check what you can implement for the benefit of the next class you are teaching. Encapsulating difficult theories in nice scientific or every-day stories of applications can make them more attractive and helps the students to understand why they are important.

If you made a mistake and students pick you up on this, openly admit the mistake and express your gratitude for pointing this out. Stay calm and cool, do not feel challenged, as everybody keeps making mistakes. Nobody is perfect, even not a professor. Remember, it is always good to see that students pay attention.

Always take students seriously, learn their names and use them, address their concerns and answer their questions. Treat them as peers and try to support their studies and their careers as much as possible by using your connections. This is an important way of passing on the kindness you have received from your mentors and encourages the students to do the same when they have a chance to do so. As part of this always write positive and constructive reference letters for your students, if they ask you to provide such letters as part of their job applications.

Never turn down students who want to see you and ask for help. However, kindly ask them to make appointments by email and to explain their concerns beforehand so that you can better plan your time and prepare. The requirement to describe their problems beforehand in an email is important, because it not only allows you to prepare but also makes the student reflect on the problem. This is a crucial part of training students to become independent: Many have found their own solutions while explaining them in an email or in person and this is in fact the best outcome a university teacher can hope for.

7.3   A Brief Word About Examinations

Prepare the students for examination and limit the areas they need to revise. Make preparation for examination not too complicated and work intensive so that they can also enjoy the process and more importantly have time to reflect on the subject matter.

As part of their preparation encourage students to go through examination questions in groups. This will reduce the chances of misinterpretation of examination questions and gives ideas for how to answer them. However, such meetings need to be complemented by individual learning.

Group work is good for more sophisticated tasks and to promote social skills. However, it can present difficulties for assessments. It is important to assign individual tasks to every group member to avoid "free rides" and to ensure just marking.

Mark student work positively, i.e. not by painstakingly trying to find mistakes but rather by identifying evidence for granting them good marks.

If you have a choice, try to design examinations and assignments in such a way that they can be assessed and marked quickly. This gives the students early returns and frees up the lecturer's time for other responsibilities.

8

Research Cooperation and Job Applications

Abstract Cooperation is more important than ever. Cooperation is crucial for building, maintaining and extending your professional networks that may give you comfort and support in difficult times of your career. Cooperation stretches from mutual authorships, occasional invitations to departmental seminars or to large research projects. Use your imagination and resourcefulness to strengthen cooperation every day.

One of the great headaches associated with academic careers is the job uncertainty. How long will I be in employment, will my new application be granted, are there any chances of prolonging the project? The positive flip side of this is that by moving on to other places you will learn new things and also introduce your methods to another group and to new colleagues. Also, it is common practice for universities to recruit senior researchers from outside the institution to diversify academic life, i.e. there is a reward for moving on in the world. Writing job applications is therefore an important part of our professional lives.

8.1 Research Cooperation

Scientific cooperation is a "must" at all levels, from faculty and university level to national and international level. This is closely connected to networking, a pre-requisite for successful research. Effective networking is an essential skill for a researcher and can contribute considerably to your career success. Networking is making, sustaining and ultimately benefiting from formal and informal connections with others within and outside your immediate field (Hopkins et al. 2020). Assign the same importance to all levels of cooperation and maintain them as best as you can. Use every opportunity to establish new ways of cooperation. As mentioned in Chap. 6, conferences, workshops and summer schools are naturally a good forum for establishing cooperation. While doing this, do not convey the impression of competitiveness, particularly at local and national level. Always be remembered for your kindness, willingness to cooperate and for sharing resources and information. Never willingly or unwillingly upset anybody. It is very easy to unintentionally annoy others simply by carelessness and we all need to watch ourselves a little.

For extending networks, meeting publication requirements and for learning new skills small publication projects involving 1–2 external partners are useful to have on a regular basis. Every little helps. For such small projects, change cooperation partners from time to time for diversity. These small projects do not necessarily need funding and you can run them in addition to ongoing projects that are funded by research councils.

Large projects involving many partners are a good opportunity to extend your networks and to engage in interesting interdisciplinary research. Often you cannot help but think that the scientific outcome of large projects does not balance the administrative effort, since they require substantial time and work in terms of proposal writing and administration. Still they provide important research income, strengthen your reputation and networks. Working as an administrator for large projects is often a full-time job.

There are also opportunities for teaching cooperation, not only through distance learning but also through video conferencing. In times of pandemics this is particularly important. Together with teaching visits by external staff this allows you to share and reduce your teaching load and thus to free up quality time for research. At the same time this is a fantastic experience for students. Such teaching cooperation often leads to research cooperation, too.

Teaching cooperation can be funded by dedicated programmes such as Socrates/Erasmus in the EU, faculty money and by specific educational programmes. Visits and meetings that are part of this cooperation can, of course, at the same time also be used for research cooperation.

Once you have made your first achievements, why not set up a website where you describe your scientific area, refer to your publications and disseminate other important information? This is a wonderful service to the scientific communities you are involved in and it helps spreading the word about your work. Your networks also benefit from this initiative. If you have developed useful analysis or modelling software, offer it for free on your website, e.g. as dedicated R packages, and never charge people for this service. You can be sure that the effort you have put into your software comes back to you in terms of citations and collaboration. A scientific blog and using `Twitter` and `Facebook` may also be a good idea for disseminating and extending your research or to communicate your research visions.

Often you may find that it is easier to cooperate with people outside rather than inside your organisation. This is a bit sad and has something to do with organisational tensions, competition, power struggles and fear. Try to be mindful of this and be understanding.

8.2   Job Applications

Always check job advertisements even if you currently have a job, e.g. on institutional websites or on https://jobrxiv.org, see also Sect. 4.1. Who knows, something interesting may be coming your way and even if you do not wish to apply, you know what is going on in your field. Keep a register/repository of "brownie points" (keynote talks, invited lectures, guest professorships etc.) and certificates to go with your application documents.

Good reasons for job applications

1. This is good training,
2. Your name and what it stands for will be promoted,
3. You assess your "market value",
4. A successful application sometimes helps to get promoted at your home university,
5. You are prepared, if something turns out badly in your current job.

Job applications need to be written carefully, because more jobs are lost by poor application than by any other cause. Often job advertisements specify the documents required in the application, but if not otherwise specified the application should include two parts: The covering letter, applying for the job, and the *curriculum vitae* (cv). The letter should be carefully worded. It should say that you wish to apply for, the position advertised and include a few of the points mentioned in the advertisement. If a particular reference number or code was given in the advertisement, quote this reference in your letter. The letter should blow your own trumpet a little, but not to the extent of bragging, and refer the reader to the attached cv and other documents (Price 1998).

The cv typically follows a table format and there are many templates available in the internet. If not otherwise stated, the cv should include the following information:

- Full name,
- Date of birth,
- Marital status,
- Schooling, further education: Places, dates, qualifications,
- Any other relevant courses attended,
- Past positions held (in chronological order); other relevant experience,
- Particular interest; extracurricular activities,
- Published papers, books.

Some employers require two or more referees they can contact to obtain information about your work performance, behaviour and other information relevant to the position. Speak to some integer people (supervisors, line managers, mentors) you have worked with in the past and ask them. if they are willing to provide a reference for you. They usually are and you can then forward their contact information (Price 1998).

In the interviews, always give your very best and convey the impression that this is your "dream job", that you are up to the challenge and that you definitely want the job. Of course, you need to be sure about this yourself before the interview starts. You can always decline a job offer later, first secure it so that you have choices to make.

Carefully check out the website and other information of the place you apply for a job. Ask around what sort of employer the new university is and check the associated ratings in the internet. Go through lists of interview questions to be well prepared. Professional unions may be able to provide you with such lists, but you can also simply google for them in the internet. Ask colleagues that you trust to give you mock interviews.

Find out who are the members of the search committee/the interview panel. Collect as much information on these people as possible. Learn their names and address them by their names. Speak to people who work at the institution where you are applying for a new job and kindly ask them to support your application. Use your networks for this purpose. Sometimes it is sufficient to arrange a meeting or to have lunch together without mentioning the job application at all.

Once a job has been offered to you, you can politely ask, if the university would consider to offer or to help with a job for your partner as well. Many academic places now offer such *dual career* opportunities.

Check carefully, whether the faculty position offered to you is permanent/temporary and whether it is fully funded or whether you need to complement your salary with external research funds. If you have a choice, prefer fully funded job opportunities, even if your research topic is popular at the moment and therefore public funding opportunities are prolific. You never know what will happen in a few years' time and competition among colleagues can be fierce, if faculty positions are not fully funded. Also check how much teaching is to be expected. Ask for written, signed agreements to be safe. Usually you can only ask for such agreements in the transition process. It is too late asking for them once you have started in the new position. Unfortunately it is not uncommon that representatives of some institutions go back on their word that they orally gave in job interviews or later state that you must have misunderstood the meaning of their word when given earlier. Also make enquiries about your future supervisor(s). Ask around before signing a new contract. Avoid superiors that enjoy power or institutional politics. Select idealists, true academics and ethically good people who can take themselves back and actively support young researchers.

When considering a new position at another university, it is good advice to collect information on the structure of the new place to understand how things work there. This will give you a good idea about your future working conditions, e.g. how much creative freedom you will have and what the weighting of different academic tasks is like. University cultures can be quite different and it is best to be prepared. Usually it takes a year or two in the new employment to fully understand what the new workplace is really like.

The most important differences in university systems are between the *institute structure* and the *department structure*. There are also hybrid types, of course.

> **Institute structure:**   Different subject areas are organised in institutes, which are self-contained units with a full professor as head and PA (personal assistant)/secretarial staff. This structure is typical of Central Europe.
> **Department structure:**   No institutes, (full) professors and lecturers represent different subject areas (one-person institutes) with hardly any support other than from project staff. Everything is organised in "pools": academic staff, admin staff, IT staff. This structure is typical of the UK, Ireland, Scandinavia and North America.

Infrastructure, funding and support is usually better in the institute structure. Academic freedom may be greater in the department structure, since it is slightly less power-driven. The department structure is preferred in many countries mostly because it is cheaper to maintain.

Some universities offer *teaching-only jobs*. If teaching is your thing, then such jobs may be great for you. Obviously they leave little or no time for research, so if you rather love research, then it is not a good idea to apply for such jobs. In this context it is also worth considering that performance reviews and promotion procedures are still notoriously bad at acknowledging teaching efforts and rather exclusively focus on research achievements, even if the job in question was mainly designed for teaching. If you carry out only teaching for an extended period, you will potentially have gaps in your publication and grant-capture records and may potentially lose contact with your research community so that it is hard to return to a position that involves research. Similar considerations apply when spending a longer period in research administration and science management with no research opportunities for yourself.

Along similar lines, if you feel like a true academic and enjoy not too applied research, consider carefully, whether you want to work for *universities of applied sciences* ("Fachhochschulen" in the Austrian, German and Swiss education systems). These universities usually have close cooperation with industry partners and their research is very applied. Also teaching loads can be considerable at such universities and you may need to share them. Problems can arise when your research and teaching qualifications surpass those of your colleagues at a university of applied sciences. If, for example, your qualifications are higher than those of many of your colleagues at such institutions, tensions can build up and as a result cooperation can be difficult.

On the other hand these universities may be a good choice for you, if you do not like writing papers and feel to be the more applied guy. However, carefully consider the advantages and disadvantages.

9

Behaviour and Disappointments

Abstract Human behaviour and interactions is a complex topic and many books have been written on this subject, particularly on the psychology of this topic. Based on my own experience in this chapter I offer a few prompts and ideas for avoiding personal clashes and for finding solutions in difficult situations. They can serve as a starting point for your own reflections and discussions with your mentor.

Behaviour and interactions with others remain a challenge throughout one's working life and to some degree they also remain a mystery. As academics we live a busy and often a competitive life, but it pays off to never stop keeping an eye on how we interact with others at work and on how we can personally contribute to a good spirit and working environment. It usually is hard if not impossible to please everybody and even more so to change others, but there is always the possibility to make changes yourself. If you learn how to cope with difficult colleagues early in your career, particularly in the competitive atmosphere of a lab or research group, you will develop valuable people-management skills that will serve you well throughout your career (Gosling and Noordam 2011).

9.1 Behaviour and Interactions

Generally make an effort to treat everybody with respect regardless of gender, academic rank or merit, age or any other distinction. Be as kind and friendly as possible in all situations, smile when addressing people and make your kindness not dependent on whether the other person is kind to you or on how you feel that particular day. The British Ecological Society (2019) stated that it helps to smile before speaking, as smiling lifts the tone of your voice and make you sound more engaging.

Relationships with people at work are mostly rewarding and this is often an important incentive for many people to continue coming to work, even if they have lost a bit the interest in the actual purpose of work or have been a little unsuccessful for a while. The people at your workplace naturally have different backgrounds, different ambitions, interests and agendas. This makes interacting with them very interesting and colourful. Many colleagues are cheerful and happy, but some may be facing domestic difficulties or are disadvantaged by an illness. In most cases you do not know about such background factors, but they all contribute to the personality of a colleague and to how this person acts at work. If someone pulls a face when you happen to meet them at the printer or in the coffee room, this often has not anything to do with you but may have entirely different reasons. Therefore take it easy and remain friendly and kind to everybody. We have all had bad days once in a while.

Employees at universities can be from all over the world and in addition research institutions often host visitors from abroad. Naturally this requires particular care in our daily interactions, as many colleagues are not familiar with our cultural background and the patterns of behaviour and emotions that come with it. Cultural and personal differences can be huge, often much greater than at workplaces outside academia, not to mention the fact that some successful researchers are mavericks and even nerds. The latter relates to the fact that successful research often requires unorthodox and independently-minded personnel.

Particularly when you move to another country, there may be local and societal habits and agendas that you are not aware of and cannot be expected to be. Host staff often tend to forget about this or feel they do not have the time to explain such things to newcomers. Unintentionally offending local researchers as a new member of staff who joined from another country can therefore easily happen and often is hard to avoid entirely. So do not be too upset when this happens even if you took great care. On the other hand, the fact that you are new and have joined from outside the country, sets you free to a certain degree and you can use this freedom to ignore local factions and emotional boundaries between groups that do not want to work with each other. You can ignore this and stay above these rivalries. For example, as an "outsider" you have not been party to past, local quarrels that may have some tradition that you do not share. As such you are free to establish relationships with anybody at the new workplace across such emotional boundaries without being hampered too much by past agendas.

It is a well-known fact that there are a lot of emotions affecting work relationships. This is, for example, demonstrated by the frequent observation that it is often easier to cooperate with researchers outside your own university and campus than it is with colleagues who physically are much closer to you which theoretically would make meetings and discussions easier. It can even happen that you have enjoyed a fruitful cooperation with colleagues from a university in another country for many years, but when you were invited to work at that university yourself and moved to that country, all this fruitful cooperation all of a sudden dried up. Apparently there is some emotional barrier related to competitiveness that seems to get in the way of cooperation within the same organisation, campus or even the same country. This is really a shame but good to be aware of so that you actively contribute to relaxing work relationships wherever you can.

It is a good recommendation to improve our skills in dealing with people and worries on a daily basis. We can also take each conversation and meeting as a small empirical field experiment in human relationships and it helps to briefly reflect on them with as much impartiality as we can muster. If you find it hard to deal with a difficult situation yourself, why not talk to your mentor or to a representative of one of the counselling services available at your campus to discuss the matter and to seek advice.

Try to establish as many relationships as possible across different subject areas and regularly meet with people within and outside your field of expertise. Do not fall into the trap of believing that only people within your own study field are important. When interacting with others always be friendly and never intentionally or carelessly ruin relationships with colleagues. Be mindful when entering conversations and try to include and link people. Leave domestic problems at home and do not allow personal circumstances to affect work relationships, although this is admittedly not always easy. Ignore inexplicable erratic behaviour of colleagues as much as you can and try to be a greater person rather than someone whose emotions get the better of them. After all, how often do we accidentally offend others ourselves?

Tip

Do not simply take every good thing that happens to you (e.g. good supervision and advice, a good mentor) for granted as if you have deserved it. You have

(continued)

achieved a lot and believe it was all your doing? Really? Think of all the coincidences and lucky circumstances, particularly where others stepped in and helped or where it seemed a miracle that you made the right decision. Or of the many instances when your partner sacrificed time so that you could pursue your research objectives. Therefore rather try to be grateful for all your blessings and cherish them as something you need to "pay back" for through kindness and support to others that are put into your care or that you work with (see also the epigraph at the beginning of this book). Imagine: If even only a few of us would try to adopt this "policy" and pass it on to others, e.g. to our students and to the early career researchers we work with, what snowball effect could this have and what big change could we make to academia!

Emailing can be a major source of misunderstandings and can lead to considerable arguments. Generally it is a poor way of communication, as emotions and intentions cannot be convincingly transmitted and are left to interpretation and guess work at the other end. Try to avoid this trap at all costs. Find a convenient time to answer all emails you get during the day, e.g. at night or early in the morning. Do this only once a day (apart from urgent exceptions) to save quality time for research and publications. Give only short, clear replies, but make sure they cannot be misunderstood. Always start and finish with a nice phrase: "I hope you are doing well.", "It was nice to see you the other day." etc. and "Thank you very much for all your efforts.", "I truly enjoy our cooperation.", "Your advice means a lot to me." etc. and sincerely mean it. Never forget to express your appreciation and try to be as polite as possible in all situations. Do not write emails when you are tired, when you feel stressed or when you are in a bad mood. If you feel angry after reading an email, do not reply instantly, but allow for things to settle. Depending on the nature of the email, perhaps even no reply at all is the best option in such a case? Otherwise answer emails in good time, follow up where there is a lack of response to your emails and be always kind and gentle.

If you have difficult, awkward or sensitive matters to discuss (that should not be on record) use the telephone or discuss the matter face to face. Otherwise write short polite emails that give the other person the freedom to deal with the matter when s/he has time to do so.

Sometimes you need to establish firm boundaries with another colleague or to simply say "no" in a polite way. This is always possible, even to express justified criticism, provided you do this in a nice and polite way. Nobody can complain about critical issues wrapped in a friendly letter or a friendly chat. However, try to keep such instances to a minimum.

In all situations, try not to get defensive, angry, intimidated, or irritated. A big part of dealing with difficult people is having confidence in your own work. Building confidence takes time, but as you start to amass a steady stream of successful publications and projects, other people's attitudes and behaviours will matter less (Gosling and Noordam 2011).

To keep emotions low and for health reasons it is advisable to keep a good work-life balance. As part of this it is a good strategy to compensate for stress at work by sport activities. Keep physical exercise going even at the worst of times and never accept excuses. This is an important part of a healthy work-life balance and is even more important in times of pandemics. Physical exercise helps you to stay fit and to settle after a long day at work. The same, of course, also applies when you are working from home.

9.2 Difficult Situations and Disappointments

Unfortunately from time to time, even in the best establishments, situations occur that we cannot simply overlook. This is always extremely sad and most regrettable, since a lot of damage is done to all parties involved and often they leave people scarred with traumas, even if they are seemingly small. Among such offensive situations there is, for example, *bullying* and *sexual harassment*.

Even if you are shocked or frustrated, at first try to address difficult situations in a relaxed, personal way, where you express your concerns to the individual in question in a gentle but firm way and try to establish a clear boundary, i.e. a line that hostilities are not allowed to overstep. At the same time keep records, particularly of email and other written communication so that they can be used as evidence should the need arise. Bullying, for example, is a frequent offence, however, it is unfortunately not easy to prove and a "paper trail" between the individuals involved may help with this.

If negative or even offensive behaviour towards you continues, the time has come to firmly stand up against unjust treatment. Consider your options carefully and discuss them with your mentor, with friends, family and other people that you can trust before deciding to stand up for your rights. Take good time to arrive at your own, personal decision and do not act on emotions. In some cases, what you have experienced may be based on a misunderstanding or is simply the result of carelessness rather than of ill intention. In most universities, dedicated services are in place to informally discuss such problems and to help you when putting your decision to action. However, careful and friendly insisting on your and other people's rights is a fundamental civic duty in support of democracy and freedom and you should not be afraid to do so.

Anyone who proposes to do good must not expect people to roll stones out of his way, but must accept his lot calmly, even if they roll a few stones upon it.

Albert Schweitzer

It also often occurs that other people at your workplace have opinions about your work performance and your duties even if not invited. Some of these comments are nevertheless useful and should not be simply dismissed or rejected. They may, of course, also be offered to you as part of an annual, formal performance review or of a forward job plan where they belong. However, even in this context, occasionally you may come across opinions that are not helpful or even hurt. You can take those up directly with your supervisor/line manager during the review session or later with your academic mentor. In most cases it does not make sense to act on faulty criticism. If, however, potential consequences may be serious, you need to consider an appropriate and well-balanced response. Again, here your mentor may be in a situation to help.

Frequently negative comments are simply an expression of unhappiness or insecurity on the side of an individual putting them forward. They may envy your success or feel somehow threatened. Unjust criticism is often a disguised compliment. No one ever kicks a dead dog (Carnegie 1998). If this is likely the case, it is best to ignore negative comments as much as possible and to remember that many great people have suffered from unjust criticism and obstacles planted in their way, for example, Johann Sebastian Bach. Here is another example (Barnes 2016, p. 125):

In his book "The Noise of Time" Julian Barnes described how the Soviet composer Dmitri Dmitrievich Shostakovich responded to great disappointment and offensive comments made by others around him, something he referred to as "noise". If you replace music by research, perhaps this will help you in difficult times:

"What could be put up against the noise of time, when there seemed to be nothing but nonsense in the world? Only the music which is inside ourselves – the music of our being – which is transformed by some into real music. Which, over the decades, if it is strong and true and pure enough to drown out the noise of time, is transformed into the whisper of history. Good music would always be good music, and great music was impregnable. This was what he held to."

Whatever fate is going to throw at you, do not let it diminish your idealism and enthusiasm. Throughout your life hold fast to the thoughts which inspire you and to your belief in the good and true. It is through our idealism that we catch sight of truth and in that idealism we possess a wealth which we must never exchange for anything else. The power of ideals is incalculable. Therefore we must grow into our ideals, so that life can never take them away from us (Schweitzer 1924).

A good way to deal with any disappointment is the strategy to turn it into something good: If you think someone was unfair to you, make a mental note and try to learn from this negative example so that you avoid doing the same or a similar thing to others. Take any injustice against yourself or others as an opportunity to learn and to improve things so that when you come to power the world around you becomes a better place. Definitely avoid adopting or copying the same bad behaviour that you have suffered from yourself. In fact this happens often enough, therefore be mindful and observe yourself and try to fight any impulse to follow negative patterns yourself.

Thoughts of revenge or punishment often follow the impulse immediately after having been offended or attacked. This is an emotional response to feeling hurt, but these thoughts are negative and destructive. Nothing good will ever come of them and they are likely to increase the feeling of being hurt. If instead you attempt to do something good, this stops the negative spiral and is in fact the best way to overcome offences.

Try to be greater than local quarrels and rise above them, definitely do not get involved in them. If they target you, make it a point that you are not interested in participating in such negative activities.

10

Sustainable Data Management

Abstract Research in natural and life sciences usually involves collecting much data and handling many files. Data are the centrepiece of our research papers and presentations. Much thought is given to experimental design (see Chap. 2) and the organisation of field work not to mention the money data collection costs. Once the data have been brought in, plausibility checks are carried out followed by the analysis and perhaps some modelling work. Then we usually write our papers and as part of them interpret and discuss the results.

But what happens to all these data and files afterwards? How do we store them and how can we make sure that we find the right files after a while and are able to remember the processes and steps leading to the figures we published? Since tax payers have paid for the collection of our research data, we are supposed to share them with others in the world and this also requires at least some basic documentation. Sustainable data management and storage are an important part of academic life and not much attention has been paid to this in the past.

10.1 Organisation and Documentation of Data

Up to now comparatively little effort is spent considering how to store and share data so that they can be re-analysed either by the same researcher or by others. Most researchers are not data or computer scientists and have received little education relating to these matters. Since everybody is under much time constraints, data management is usually one of the last things people tend

© The Author(s), under exclusive license to Springer Nature Switzerland AG 2021
A. Pommerening, *Staying on Top in Academia*,
https://doi.org/10.1007/978-3-030-65467-2_10

to think of. Increasingly, however, it is important to store and document scientific data in ways that facilitate Open Science and the effective retrieval and interpretation of data in the future (Borer et al. 2009). Storing research data has also become a legal requirement in many countries, for example, by means of the Freedom of Information Act in the UK. This implies that the responsible research institution must manage their data in a form that it can be made available to anyone making a legitimate request to the institution. There are only a few grounds on which such a request can be legally refused (such as commercial or research subject confidentiality) but the rules are quite strict. In addition many funding bodies even make it a condition of all of their funding that resulting data must be made available to all through a readily accessible (web-based) portal.

According to the British Ecological Society (2014) research data are the factual pieces of information used to produce and validate research results and can be classified into five categories:

- **Observational**: Data are tied to time and place and are irreplaceable (e.g. field observations, weather station readings, satellite data),
- **Experimental**: Data generated in a controlled or partially controlled environment which can be reproduced, although it may be expensive to do so (e.g. field plots or greenhouse experiments, chemical analyses),
- **Simulation**: Data generated from models (e.g. climate or population modelling),
- **Derived**: Data which are not collected directly but generated from (an)other data file(s), e.g. a population biomass which has been calculated from population density and average body size data,
- **Metadata**: Data about data providing contextual information as explained in this chapter.

Data typically have a longer lifespan than the project they were created for. Some projects may only focus on certain parts of the lifecycle, such as primary data creation or reusing others' data. Traditionally researchers were mainly concerned with the early stages of the lifecycle, i.e. creating, processing and using. Now a combination of technologies that allow data sharing and the increasing need to combine different datasets to address various research questions means that preserving and sharing data has become an important part of the scientific process (British Ecological Society 2014).

Wilkinson et al. (2016) published the *FAIR Guiding Principles* for scientific data management and stewardship, see https://www.go-fair.org/fair-principles/. FAIR is an acronym for *Findability, Accessibility, Interoperability* and *Reuse*. Findability relates to the requirement that data can be identified easily. Once the user has found the data, s/he needs to know how to access the data including authentication and authorisation. Data often interoperate with other data, with applications or specific workflows for analysis, storage and processing. The ultimate objective of FAIR is to optimise the reuse of data which requires good documentation.

In the past, much data analysis and modelling was carried out in spreadsheet software including a number of worksheets and many tables in each of them. This organisation of work has turned out to be confusing and after a short while it is hard to tell how exactly the data were analysed and in which order. Many researchers have then faced real difficulties when trying to reconstruct the results they had arrived at earlier and particular the graphs. Returning to earlier analysis steps is often necessary as part of the review process, when revisions of figures or tables are requested. Most analyses are re-run many times before they are finished, therefore the smoother and more automated the workflow, the easier, faster and more robust the process of repeating it will be (Grimm et al. 2014; British Ecological Society 2018).

A particular problem when revisiting data is the missing documentation of the data including *metadata* (e.g. containing soil, climate, other environmental and management data) and the *workflow* (i.e. the individual steps taken and their order) applied in the analyses. Apart from the fact that this creates personal problems and delays work, it is not possible to share the data with the research community. Missing documentation therefore leads to a situation with no *reproducibility*, no *transparency* and no potential *reuse* (Strasser et al. 2011). Producing good documentation and metadata ensures that data can be understood and used in the long term (British Ecological Society 2014).

The easiest way to achieve some level of documentation is to carry out the data analysis with *scripted analysis software* such as Julia, MATLAB, Python, R or SAS rather than software that is driven by graphical user interfaces (GUI). GUI-based analyses often seem convenient when you are pushed for time and are rapidly working through your data and analysis, but they rarely give a clear account of what exactly you have done (Strasser et al. 2011). By contrast, analysis scripts serve as written records of the various steps involved in processing and analysing

(continued)

data and to some degree they also provide a form of analytical metadata. Scripted analysis code also has the advantage that all your analysis is eventually automated. Often enough mistakes are introduced by manual work, particularly when you are tired or are doing routine work many times over. Automated scripts ensure that all repetitions of the analysis are carried out with the same precision and avoid human error.

Another good suggestion is to use descriptive, self-explanatory and succinct file and variable names (using the English language) that give you clues about the objects studied (e.g. the type of plants), the year of collection and perhaps the number of the research plot, e.g. `Trees_ArtistsWood_2011_Plot1.xls` or `TreesArtistsWood2011 Plot1.xls` (Strasser et al. 2011). The names should be unique and reflect the contents of the file. They should avoid spaces, punctuation, accented characters. More specifically they should stick to "a-zA-Z0-9" characters (British Ecological Society 2018). This informative naming convention will make it easier for you to remember the way round your data, even if a few months or years have passed, since you used them last.

In a similar way store any metadata in separate files. Use similar naming conventions and keeping one column with a common identifier that you also used in related files, so that you can automatically merge or join the data from two or more files in `Julia`, `MATLAB`, `Python`, `R` or `SAS`. For example, site and plot name are possible identifiers that could be used in all files that are connected.

From time to time it happens that software companies go out of business. Then their software products are not maintained any more and after a while become unavailable or do not run any more when you buy and use a new computer. Also, it can happen that software is maintained, however, the rapid process of continued software updating has created incompatibilities with files you saved in earlier versions of this software. If that happens, your data can disappear and this can seriously stall your research progress. A good solution can be to store data as far as possible in *non-proprietary software* based on open formats, e.g. using tab delimited text files (extension `*.txt`) or comma delimited text files (extension `*.csv`), as text files can always be read. Other non-proprietary formats include `GIF`, `JPEG` and `PNG` for images.

Always keep an *uncorrected original data file* that includes all possible measurement and typing errors and do not make any corrections to this file.

When you make corrections to an original file, you could easily be changing something that you later discover was correct in its original form or you introduced another mistake while correcting the file (Strasser et al. 2011). Instead carry out corrections, cleaning, merging, transforming etc. within your analysis script (e.g. in Julia, MATLAB, Python, R or SAS) and do not do any of this manually. Using a scripted language, you can re-run the analyses as well as transformations and corrections to your data with your original data as input, but saving the changes to a separate data file.

We rarely document the way how we get from the raw data to the final results of our research, i.e. the *workflow*. This could be achieved by drawing a simple flow chart, e.g. as part of our research plan, see Sect. 5.4. Another way of documenting the workflow is simply to comment our analysis scripts in Julia, MATLAB, Python, R or SAS. Well-documented code is easier to review, to share with others inside and outside our groups and to use in repeated analyses. Documented workflows ensure *reproducibility, transparency* and *reuse* which are important requirements of modern research.

Result and manuscript files can be kept in folders that include different dates in the file names so that the folder and file structures offer a form of basic version control. In a similar way all results and manuscript files could be stored in one folder but contained multiple versions of the same files with different dates included in the file names, e.g. an earlier version of the manuscript SpatialSpeciesDiversity03052020.docx from 3 May 2020 and a later version SpatialSpeciesDiversity08092020.docx from 8 September of the same year. This ensures that you have copies of all (important) earlier versions of analysis scripts, result and manuscript files safely stored, as it may later turn out that you deleted important parts or made mistakes and for correcting need to go back to earlier versions. Instead of creating separate, dated files for each day that you are working on a project, you could limit your effort to variants of your files that mark big changes. For example, you could create a new manuscript copy and assign a date to the file name *before* you completely re-work the structure of your article. Some researchers recommend using strict *version control* for analysis scripts as well as data files and manuscripts so that even the smallest change is saved. This can, for example, be achieved using GitHub (https://github.com/) and similar repositories such as GitLab, Bitbucket or Savannah. However, in my opinion this is not necessary and a simpler variant as described in this paragraph is sufficient. Incidentally, GitHub and similar software do not have to be used for research purposes only. They were originally developed for collaborative software development and are also excellent repositories for teaching material (not too dissimilar to commercial teaching platforms such as `Blackboard`, `Brightspace`, `Canvas` and `Moodle`) where simple documents can be prepared using the `Markdown` language.

Some researchers have one main or top folder for each publication project. To keep track of data and know how to find them, digital files and folders should be structured and well-organised. This is particularly important when working in collaboration (British Ecological Society 2014). However, there is no single best way to organise a file system. The key is to make sure that the structure of directories and location of files is consistent, informative and works for you. An example of a possible basic project directory structure is (British Ecological Society 2018):

- The `data` folder contains all input data (and metadata) used in the analysis. You may want to have a sub-folder entitled `cleaned_data` or `final_data`. `README.txt` files mentioned in Sect. 10.2 could go here.
- The `doc` folder includes all data and code documentation. This could include any rationale, concerns, ideas etc.
- The `paper` folder contains the manuscript, figures and tables and sub-folders could be made for the different journals you attempted including revisions.
- The analysis and modelling results are stored in the folder `results`.
- A dedicated directory `R` or `SAS` collects all scripts used in the analysis/modelling.
- The `reports` folder contains intermediate reports and notes that you have written during the conceptual work, the analysis and modelling. This folder can also include the research plan, see Sect. 5.4. You can also store important email communication here or in a separate folder.

Needless to say that any data on your computer's hard drives are not safe. Therefore you need to design *backup strategies* right from the start. You should have several copies of your data on separate external drives (such as USB sticks or external hard drives), in safe clouds and possibly on CDs and DVDs. You also should consider that clouds are synchronised in different ways and that you may need to choose a synchronisation scheme that works best for your objectives and data structure. Usually the academic institution you work for has also limited server space available to you to store important data. Although university servers are protected by a number of technical routines (e.g. backup services), any of these data storage devices can go off any time and for data sustainability everybody is well advised to use a number of different personal storage solutions to be on the safe side. You also need to consider migrating data files from time to time, since storage media and hard drives can degrade over time or become outdated (British Ecological Society 2014).

As an experienced computer user you are certainly aware of *computer-virus attacks* and the multiple, canning strategies that malicious companies and individuals have adopted to penetrate and destabilise our computer systems. These can seriously jeopardise our research data. Most of these attacks are related to emailing, however, there are also other entry points. The recent increase of *ransomware attacks*, for example, suggests that you should not entirely rely on clouds and servers for data storage, but also keep copies of your data on external drives etc. offline. Discussing these serious issues in detail is beyond the scope of this book and you are strongly advised to consult your institutional IT department and the services they offer.

10.2 Publications and Data Archives

Many scientific journals and funding bodies now require the data used in the analyses of the papers they publish or they have funded to be made publicly available in *data archives* or *research data repositories*. Some funders even require *data managing* and *sharing plans* as part of grant applications (British Ecological Society 2014), see Sect. 5.10. Data archiving is the long term storage of data and methods. This requirement clearly is a laudable move deserving our support and again facilitates important requirements of modern research. i.e. reproducibility, transparency and reuse. There are currently no generally accepted standards for where and how data and methods should be published (British Ecological Society 2018).

Data archiving and publishing, however, also promotes data sharing. This is the process of making information, particularly data generated from research, available to all. Data sharing is an important requirement of modern research and, as stated earlier, supports Open Science and the initiation of new research. It also promotes data longevity, i.e. that collected data (that usually are paid for by tax payers) continue to exist and to be useful to future generations of researchers and can be reused multiple times so that value is continuously added to the effort of collecting them.

By sharing data you expose them to other potential projects and collaborators. Sharing data can be daunting, but data are valuable resources and their usefulness could extend far beyond the original purpose for which they were created (British Ecological Society 2014). Sharing also potentially increases the citation of source papers, i.e. your papers where the data were used for the first time. A similar effect is achieved by making any analysis scripts or packages that you used in your papers available to the international research community. Sharing data also allows others to reproduce and confirm your findings which again contributes to transparency (Ritchie 2020). Finally, data sharing reduces the costs of duplicating data collection and advances science by letting others use data in innovative ways. Exceptions may include records containing sensitive information about endangered or threatened species and personal information legally protected by data protection acts.

Also for sharing data a basic documentation defining the contents of your data files is useful and necessary. You need to prepare one anyway for your own sake as discussed in Sect. 10.1. Such documentation could include the project name, project summary, funding information, primary contact information, supporting information, dates of data collection, version, units, parameter names and formats. This basic information could be stored in a 'read me' text file, i.e. `README.txt`. All information, names and descriptors should naturally be based on the English language. The organisation of the data including file structure and formats should be consistent. Again, descriptive file names are of great help here and basic plausibility checks (checking questionable or impossible values and outliers, which may just be typos from data entry) ensuring quality should be included.

There are many descriptions and recommendations for data repositories and archives available in the internet. Many universities offer their own institutional archives that all research staff can use for free.

> `Dryad`, for example, is a general data repository mainly used for life science data. The `Open Science Framework` is a general data archive. `Figshare` is a general data repository for a wide range of research outputs including data sets. `Zenodo` is a data archive specialising in archiving and sharing computer code. `Code Ocean` is a cloud-based platform for sharing, discovering and running code. Finally `Harvard Dataverse` and `Mendeley Data` are general data repositories.

Some of these services are free, others are commercial. Software and data deposited with any of these services are provided with a globally unique Digital Object Identifier (DOI) and storage is guaranteed for a long period of time. Some of these archives can be linked with personal `GitHub` accounts, so that you can deposit your research data and methods in `GitHub` and authorship and software version are automatically recorded by the aforementioned repositories (British Ecological Society 2018).

References

Barnes, J. (2016). *The noise of time* (184p). London: Vintage.

Biotechnology and Biological Sciences Research Council. (2016). *Academic career mentoring and best practice for formal mentoring programmes* (9p). London.

Borer, E. T., Seabloom, E. W., Jones, M. B., & Schildhauer, M. (2009). Some simple guidelines for effective data management. *Bulletin of the Ecological Society of America, 90*, 205–214.

British Ecological Society. (2014). *A guide to data management in ecology and evolution* (21p). London.

British Ecological Society. (2015). *A guide to getting published in ecology and evolution* (40p). London.

British Ecological Society. (2018). *A guide to reproducible code in ecology and evolution* (44p). London.

British Ecological Society. (2019). *Promoting your research* (41p). London.

Biggs, J., & Tang, C. (2011). *Teaching for quality learning at university* (4nd ed., 4187p). Maidenhead: Open University Press.

Carnegie, D. (1998). *How to stop worrying and start living* (304p). London: Vermillion.

Cornér, S., Löfström, E., & Pyhälto, K. (2017). The relationship between doctoral students' perceptions of supervision and burnout. *International Journal of Doctoral Studies, 12*, 91–106.

Crawley, M. J. (2005). *Statistics. An introduction using R* (327p). Chichester: Wiley.

Davies, S. (2007). *The Mabinogi. Translated with an introduction and notes by Sioned Davies* (293p). Oxford: Oxford University Press.

Fanelli, D. (2010). Do pressures to publish increase scientists' bias? An empirical support from US States Data. *PLoS ONE, 5*, e10271.

Gadow, K. v., & Bredenkamp, B. (1992). *Forest management* (151p). Pretoria: Academia.

Gosling, P., & Noordam, B. (2011). *Mastering your PhD* (2nd ed., 237p). Dordrecht: Springer.

Grimm, V., Berger, U., DeAngelis, D. L., Polhill, J. G., Giske, J., & Railsback, S. F. (2010). The ODD protocol: A review and first update. *Ecological Modelling, 221,* 2760–2768.

Grimm, V., Augusiak, J., Focks, A., Frank, B. M., Gabsi, F., Johnston, A. S. A., Liu, C., Martin, B. T., Meli, M., Radchuk, V., Thorbeck, P., & Railsback, S. F. (2014). Towards better modelling and decision support: Documenting model development, testing, and analysis using TRACE. *Ecological Modelling, 280,* 129–139.

Hirsch, J. E. (2005). An index to quantify an individual's scientific research output. *Proceedings of the National Academy of Sciences of the United States of America, 102,* 16569–16572.

Hopkins, S., Brooks, S. A., & Yeung, A. (2020). *Mentoring to empower researchers* (220p). London: SAGE Publications.

Hotaling, S. (2020). Simple rules for concise scientific writing. *Limnology and Oceanography Letters, 2,* 10165.

Kimmins, J. P. (2004). *Forest ecology – A foundation for sustainable management* (3rd ed., 700p). Upper Saddle River: Pearson Education Prentice Hall.

Lewis, G., & Williams, R. (2019). *The book of Taliesin. Poems of warfare and praise in an enchanted Britain* (224p). London: Penguin Classics.

Livoreil, B., Glanville, J., Haddaway, N. R., Bayliss, H., Bethel, A., de Lachapelle, F. F., Robalino, S., Savilaakso, S., Zhou, W. T., Petrokofsky, G., & Frampton, G. (2017). Systematic searching for environmental evidence using multiple tools and sources. *Environmental Evidence, 6,* 23.

Matosin, N., Frank, E., Engel, M., Lum, J. S., & Newell, K. A. (2014). Negativity towards negative results: A discussion of the disconnect between scientific worth and scientific culture. *Disease Models & Mechanisms, 7,* 171–173.

Molina Jordá, J. (2013). Academic tutoring at the univeersity level: Development and promotion methodology through project work. *Procedia – Social and Behavioral Sciences, 106,* 2594–2601.

Montgomery, D. C. (2013). *Design and analysis of experiments* (8th ed., 726p). New Delhi: Wiley.

Morrison, L. J., Lorens, E., Bandiera, G., Lilies, W. C., Lee, L., Hyland, R. A., McDonald-Blumer, H., Allard, J. P., Panisko, D. M., Heathcote, E. J., & Levinson, W. (2014). Impact of a formal mentoring program on academic promotion of Department of Medicine faculty: A comparative study. *Medical Teacher, 36,* 608–614.

Newton, A. C. (2007). *Forest ecology and conservation. A handbook of techniques* (454p). Oxford: Oxford University Press.

Nouri, J. (2016). The flipped classroom: For active, effective and increased learning – Especially for low achievers. *International Journal of Educational Technology in Higher Education, 13,* 33.

Porkress, R. (2004). *Collins dictionary statistics* (316p). Glasgow: Harper Collins Publishers.

Price, C. (1998). *The style and presentation of written work* (Rev. ed., 62p). Bangor: Bangor University.

Ritchie, S. (2020). *Science fictions. Exposing fraud, bias, negligence and hype in science* (353p). London: The Bodley Head.

Schweitzer, A. (1924). *Memoirs of childhood and youth* (124p). London: The Macmillan Company.

Speidel, G. (1972). *Planung im Forstbetrieb. Grundlagen und Methoden der Forstein-richtung.* [*Planning in a forest enterprise. Basics and Methods of forest planning.*] (267p). Berlin: Parey Buchverlag.

Strasser, C., Cook, R., Michener, W., & Budden, A. (2011). DataOne – Primer on data management: What you always wanted to know (11p). www.dataone.org.

Sutherland, W. J., Pullin, A. S., Dolman, P. M., & Knight, T. M. (2004). The need for evidence-based conservation. *Trends in Ecology and Evolution, 19*, 305–308.

Tomas, L., Evans, N., Doyle, T., & Skamp, K. (2019). Are first year students ready for a flipped classroom? A case for a flipped learning continuum. *International Journal of Educational Technology in Higher Education, 16*, 5.

Trinity Counselling. (2020). *Trinity career mentor handbook (2019–2020)* (39p). Dublin: Trinity College Dublin, The University of Dublin.

Wilkinson, M. D., Dumontier, M., Aalbersberg, I. J., Appleton, G., Axton, M., Baak, A., Blomberg, N., Boiten, J.-W., Bonino da Silva Santos, L., Bourne, P. E., Bouwman, J., Brookes, A. J., Clark, T., Crosas, M., Dillo, I., Dumon, O., Edmunds, S., Evelo, C. T., Finkers, R., Gonzalez-Beltran, A., Gray, A. J. G., Groth, P., Goble, C., Grethe, J. S., Heringa, J., 't Hoen, P. A. C., Hooft, R., Kuhn, T., Kok, R., Kok, J., Lusher, S. J., Martone, M. E., Mons, A., Packer, A. L., Persson, B., Rocca-Serra, P., Roos, M., van Schaik, R., Sansone, S.-A., Schultes, E., Sengstag, T., Slater, T., Strawn, G., Swertz, M. A., Thompson, M., van der Lei, J., van Mulligen, E., Velterop, J., Waagmeester, A., Wittenburg, P., Wolstencroft, P., Zhao, J., & Mons, B. (2016). The FAIR Guiding Principles for scientific data management and stewardship. *Scientific Data, 3*, 160018.

Index

Printed in the United States
by Baker & Taylor Publisher Services